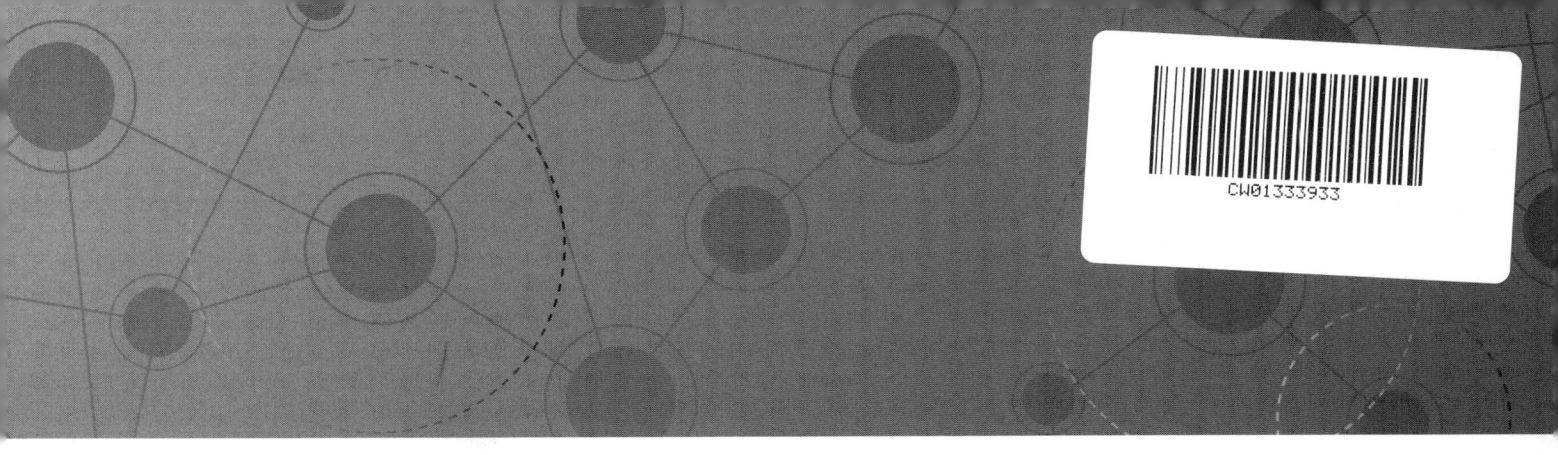

CSX™ Cybersecurity Fundamentals

Study Guide, 2nd Edition

ISACA
Trust in, and value from, information systems

www.isaca.org/cyber

ISACA®

ISACA *(isaca.org)* helps global professionals lead, adapt and assure trust in an evolving digital world by offering innovative and world-class knowledge, standards, networking, credentialing and career development. Established in 1969, ISACA is a global nonprofit association of 140,000 professionals in 180 countries. ISACA also offers the Cybersecurity Nexus™ (CSX), a holistic cybersecurity resource, and COBIT®, a business framework to govern enterprise technology.

Disclaimer

ISACA has designed and created *Cybersecurity Fundamentals Study Guide, 2nd Edition* primarily as an educational resource for cybersecurity professionals. ISACA makes no claim, representation or warranty that use of any of this study guide will assure a successful outcome or result in any certificate or certification. The study guide was produced independently from the CSX Fundamentals exam. Copies of current or past exams are not released to the public and were not used in the preparation of this publication.

Reservation of Rights

© 2017 ISACA. All rights reserved. No part of this publication may be used, copied, reproduced, modified, distributed, displayed, stored in a retrieval system or transmitted in any form by any means (electronic, mechanical, photocopying, recording or otherwise), without the prior written authorization of ISACA.

ISACA
3701 Algonquin Road, Suite 1010
Rolling Meadows, IL 60008 USA
Phone: +1.847.253.1545
Fax: +1.847.253.1443
Email: info@isaca.org
Website: *www.isaca.org*

Provide Feedback: *www.isaca.org/cyber-fundamentals-study-guide*
Participate in the ISACA Knowledge Center: *www.isaca.org/knowledge-center*
Follow ISACA on Twitter: *https://twitter.com/ISACANews*
Join ISACA on LinkedIn: ISACA (Official), *http://linkd.in/ISACAOfficial*
Like ISACA on Facebook: *www.facebook.com/ISACAHQ*

Cybersecurity Fundamentals Study Guide, 2nd Edition
ISBN 978-1-60420-699-9

ACKNOWLEDGMENTS

The *Cybersecurity Fundamentals Study Guide, 2nd Edition* development is the result of the collective efforts of many volunteers. ISACA members from throughout the world participated, generously offering their talent and expertise.

Special thanks go to Patric J.M. Versteeg, CISA, CISM, CRISC, CGEIT, CSX-P, VSec, The Netherlands, who served as lead subject matter reviewer.

Expert Reviewers
Gurvinder P. Singh, CISA, CISM, CRISC, Sydney Trains, Australia
John Tannahill, CISM, CGEIT, CRISC, J. Tannahill & Associates, Canada
Balasubramaniyan Pandian, CISSP, ISO27kLA, Triquesta, Singapore
KyoungGon Kim, CISA, CISSP, Deloitte, South Korea
Derek Grocke, HAMBS, Australia
Vilius Benetis, CISA, CRISC, NRD CS, Lithuania
Alberto Ramirez Ayon, CISA, CISM, CRISC, CBCP, CIAM, Seguros Monterrey New York Life, Mexico
Matthew Morin, Cylance, Inc., USA

ISACA Board of Directors
Christos K. Dimitriadis, Ph.D., CISA, CISM, CRISC, INTRALOT S.A., Greece, Chair
Theresa Grafenstine, CISA, CGEIT, CRISC, CIA, CGAP, CGMA, CPA, U.S. House of Representatives, USA, Vice-chair
Robert Clyde, CISM, Clyde Consulting LLC, USA, Director
Leonard Ong, CISA, CISM, CGEIT, CRISC, CPP, CFE, PMP, CIPM, CIPT, CISSP ISSMP-ISSAP, CSSLP, CITBCM, GCIA, GCIH, GSNA, GCFA, Merck, Singapore, Director
Andre Pitkowski, CGEIT, CRISC, OCTAVE, CRMA, ISO27kLA, ISO31kLA, APIT Consultoria de Informatica Ltd., Brazil, Director
Eddie Schwartz, CISA, CISM, CISSP-ISSEP, PMP, USA, Director
Jo Stewart-Rattray, CISA, CISM, CGEIT, CRISC, FACS CP, BRM Holdich, Australia, Director
Tichaona Zororo, CISA, CISM, CGEIT, CRISC, CIA, CRMA, EGIT | Enterprise Governance (Pty) Ltd., South Africa, Director
Zubin Chagpar, CISA, CISM, PMP, Amazon Web Services, UK, Director
Rajaramiyer Venketaramani Raghu, CISA, CRISC, Versatilist Consulting India Pvt. Ltd., India, Director
Jeff Spivey, CRISC, CPP, Security Risk Management, Inc., USA, Director
Robert E Stroud, CGEIT, CRISC, Forrester Research, USA, Past Chair
Tony Hayes, CGEIT, AFCHSE, CHE, FACS, FCPA, FIIA, Queensland Government, Australia, Past Chair
Greg Grocholski, CISA, SABIC, Saudi Arabia, Past Chair
Matt Loeb, CGEIT, FASAE, CAE, ISACA, USA, Director

Cybersecurity Working Group
Eddie Schwartz, CISA, CISM, CISSP-ISSEP, PMP, USA, Chair
Niall Casey, Johnson & Johnson, USA
Stacey Halota, CISA, CISSP and CIPP, Graham Holdings, USA
Tammy Moskites, CISM, Venafi, USA
Lisa O'Connor, Accenture, USA
Ron Ritchey, JPMorgan Chase & Co., USA
Marcus Sachs, North American Electric Reliability Corporation, USA
Greg Witte, CISM, CISSP-ISSEP, PMP, G2, Inc., USA
Rogerio Winter, Brazilian Army, Brazil

Special Recognition for Financial Suppport
ISACA New Jersey Chapter

Page intentionally left blank

CONTENTS

Section 1: Cybersecurity Introduction and Overview 3
    Topic 1—Introduction to Cybersecurity 5
    Topic 2—Difference Between Information Security and Cybersecurity 11
    Topic 3—Cybersecurity Objectives 13
    Topic 4—Cybersecurity Governance 15
    Topic 5—Cybersecurity Domains 19
    Section 1—Knowledge Check 21

Section 2: Cybersecurity Concepts 23
    Topic 1—Risk 25
    Topic 2—Common Attack Types and Vectors 33
    Topic 3—Policies 39
    Topic 4—Cybersecurity Controls 45
    Section 2—Knowledge Check 48

Section 3: Security Architecture Principles 49
    Topic 1—Overview of Security Architecture 51
    Topic 2—The OSI Model 55
    Topic 3—Defense in Depth 59
    Topic 4—Information Flow Control 61
    Topic 5—Isolation and Segmentation 68
    Topic 6—Logging, Monitoring and Detection 71
    Topic 7—Encryption Fundamentals, Techniques and Applications 75
    Section 3—Knowledge Check 83

Section 4: Security of Networks, Systems, Applications and Data 85
    Topic 1—Process Controls—Risk Assessments 87
    Topic 2—Process Controls—Vulnerability Management 91
    Topic 3—Process Controls—Penetration Testing 93
    Topic 4—Network Security 97
    Topic 5—Operating System Security 105
    Topic 6—Application Security 111
    Topic 7—Data Security 115
    Section 4—Knowledge Check 118

Section 5: Incident Response 119
    Topic 1—Event vs. Incident 121
    Topic 2—Security Incident Response 125
    Topic 3—Investigations, Legal Holds and Preservation 127
    Topic 4—Forensics 129
    Topic 5—Disaster Recovery and Business Continuity Plans 133
    Section 5—Knowledge Check 137

Section 6: Security Implications and Adoption of Evolving Technology 139
    Topic 1—Current Threat Landscape 141
    Topic 2—Advanced Persistent Threats 143
    Topic 3—Mobile Technology—Vulnerabilities, Threats and Risk 147
    Topic 4—Consumerization of IT and Mobile Devices 153
    Topic 5—Cloud and Digital Collaboration 157
    Section 6—Knowledge Check 161

Appendix A—Knowledge Statements 165

Appendix B—Glossary 167

Appendix C—Knowledge Check Answers 191

CYBERSECURITY FUNDAMENTALS STUDY GUIDE

Why become a cybersecurity professional? The protection of information is a critical function for all enterprises, industries and modern societies. Cybersecurity is a growing and rapidly changing field, and it is crucial that the central concepts that frame and define this increasingly pervasive field are understood by professionals who are involved and concerned with the security implications of information technology (IT). The *Cybersecurity Fundamentals Study Guide, 2nd Edition* is designed for this purpose, as well as to provide insight into the importance of cybersecurity, and the integral role of cybersecurity professionals. This guide will also cover five key areas of cybersecurity: 1) cybersecurity concepts, 2) security architecture principles, 3) security of networks, systems, applications and data, 4) incident response, and 5) the security implications of the adoption of emerging technologies.

Upon completion of this guide, the learner will be able to:
- Understand basic cybersecurity concepts and definitions.
- Understand basic risk management and risk assessment principles relating to cybersecurity threats
- Apply security architecture principles.
- Identify components of a security architecture.
- Define network security architecture concepts.
- Understand malware analysis concepts and methodology.
- Recognize the methodologies and techniques for detecting host- and network-based intrusions via intrusion detection technologies.
- Identify vulnerability assessment tools, including open source tools and their capabilities.
- Understand system hardening.
- Understand penetration testing principles, tools and techniques.
- Define network systems management principles, models, methods and tools.
- Understand remote access technology and systems administration concepts.
- Distinguish system and application security threats and vulnerabilities.
- Recognize system life cycle management principles, including software security and usability.
- Define types of incidents (categories, responses and time lines for responses).
- Outline disaster recovery and business continuity planning.
- Understand incident response and handling methodologies.
- Understand security event correlation tools and how different file types can be used for atypical behavior.
- Recognize investigative implications of hardware, operating systems and network technologies.
- Be aware of the basic concepts, practices, tools, tactics, techniques and procedures for processing digital forensic data.
- Identify network traffic analysis methods.
- Recognize new and emerging information technology and information security technologies.

Page intentionally left blank

Section 1: Cybersecurity Introduction and Overview

Topics covered in this section include:
1. Introduction to cybersecurity
2. Difference between information security and cybersecurity
3. Cybersecurity objectives
4. Cybersecurity governance
5. Cybersecurity domains

Page intentionally left blank

TOPIC 1—INTRODUCTION TO CYBERSECURITY

THE EVOLUTION OF CYBERSECURITY

Safeguarding information has been a priority for as long as people have needed to keep information secure and private. Even simple encryption techniques such as Caesar ciphers were created to ensure confidentiality. But as time and technology move forward, so do the demands of security. Today, the objective of information security is threefold, involving the critical components of confidentiality, integrity and availability (see **figure 1.1**.). All three components are concerned with the protection of information. **Confidentiality** means protection from unauthorized access, while **integrity** means protection from unauthorized modification, and **availability** means protection from disruptions in access.

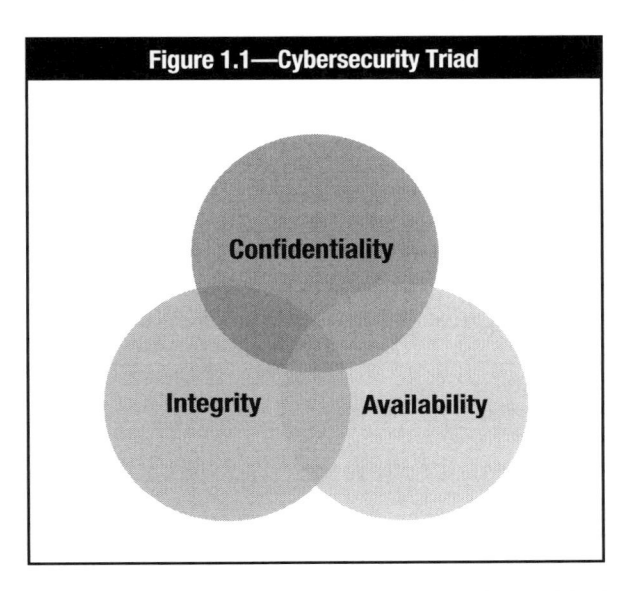

Figure 1.1—Cybersecurity Triad

- Confidentiality
- Integrity
- Availability

The terms "cybersecurity" and "information security" are often used interchangeably, but in reality, cybersecurity is a part of information security. More specifically, **cybersecurity** can be defined as the protection of information assets by addressing threats to information processed, stored and transported by internetworked information systems.

Cybersecurity usually relates to an entity initiating threats due to the existence of a global cyberspace (i.e., Internet). Unlike information security, cybersecurity does not include natural hazards, personal mistakes or physical security. To put it even simpler, if we remove offensive and adversary human behavior threats coming through interconnected systems, cybersecurity would not be an issue, and information security alone would be sufficient.

Figure 1.2 shows the complex relationship among cybersecurity and other security domains, as described in International Organization for Standardization (ISO) 27032. For example, not all critical infrastructure services (e.g., water, transportation) will directly or significantly impact the state of cybersecurity within an organization. However, a lack of proper cybersecurity measures can negatively impact the availability and reliability of the critical infrastructure systems that are used by the providers of these services (e.g., telecommunications).[1]

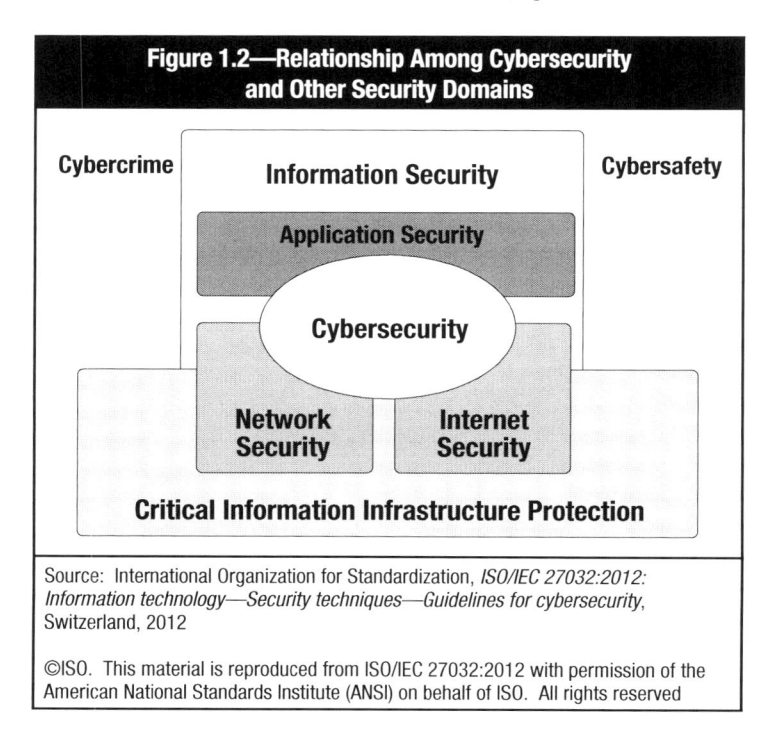

Figure 1.2—Relationship Among Cybersecurity and Other Security Domains

- Cybercrime
- Information Security
- Application Security
- Cybersecurity
- Cybersafety
- Network Security
- Internet Security
- Critical Information Infrastructure Protection

Source: International Organization for Standardization, *ISO/IEC 27032:2012: Information technology—Security techniques—Guidelines for cybersecurity*, Switzerland, 2012

©ISO. This material is reproduced from ISO/IEC 27032:2012 with permission of the American National Standards Institute (ANSI) on behalf of ISO. All rights reserved

Managing cybersecurity issues requires coordination between many entities—public and private, local and globally—as cybersecurity is closely tied to the security of the Internet, enterprise and home networks, and information security. This can be complicated because, due to matters of national security, some critical infrastructure services are not openly discussed and knowledge of weaknesses to these services can have a direct impact on security. Therefore, a basic framework for information sharing and incident coordination is needed to provide assurance to stakeholders that cybersecurity issues are being addressed.[2]

CYBERSECURITY AND SITUATIONAL AWARENESS

Cybersecurity plays a significant role in the ever-evolving cyber landscape. New trends in mobility and connectivity present a broad range of challenges as new attacks continue to develop along with emerging technologies. Cybersecurity professionals must be informed and flexible to identify and manage potential new threats, such as new cybercrime methods and advanced persistent threats (APTs), effectively. APTs are attacks by an adversary who possesses sophisticated levels of expertise and has the time, patience and significant resources, which allow the attacker to create opportunities to achieve its objectives using multiple attack vectors.

To successfully protect their systems and information, cybersecurity professionals must demonstrate a high degree of situational awareness. This takes time to cultivate, because it usually develops through experience within a specific organization. Each organization has its own distinct culture, which means that conditions vary widely from one organization to another. Therefore, it is critical for cybersecurity professionals to have a detailed understanding of the environment in which they operate and a broader ongoing awareness of threats affecting other organizations and industries to ensure relevant knowledge is maintained.

---

[1] International Organization for Standardization, *ISO/IEC 27032:2012: Information technology—Security techniques—Guidelines for cybersecurity*, Switzerland, 2012
[2] *Ibid.*

Section 1: Cybersecurity Introduction and Overview

The protection of assets from threats is a primary concern for security in general. Threats, in turn, are then categorized related to the likelihood (i.e., potential) that they will impact protected assets. In security, threats that are related to malicious or other human activities are often given greater attention. **Figure 1.3** gives an illustration of these security concepts and relationships.

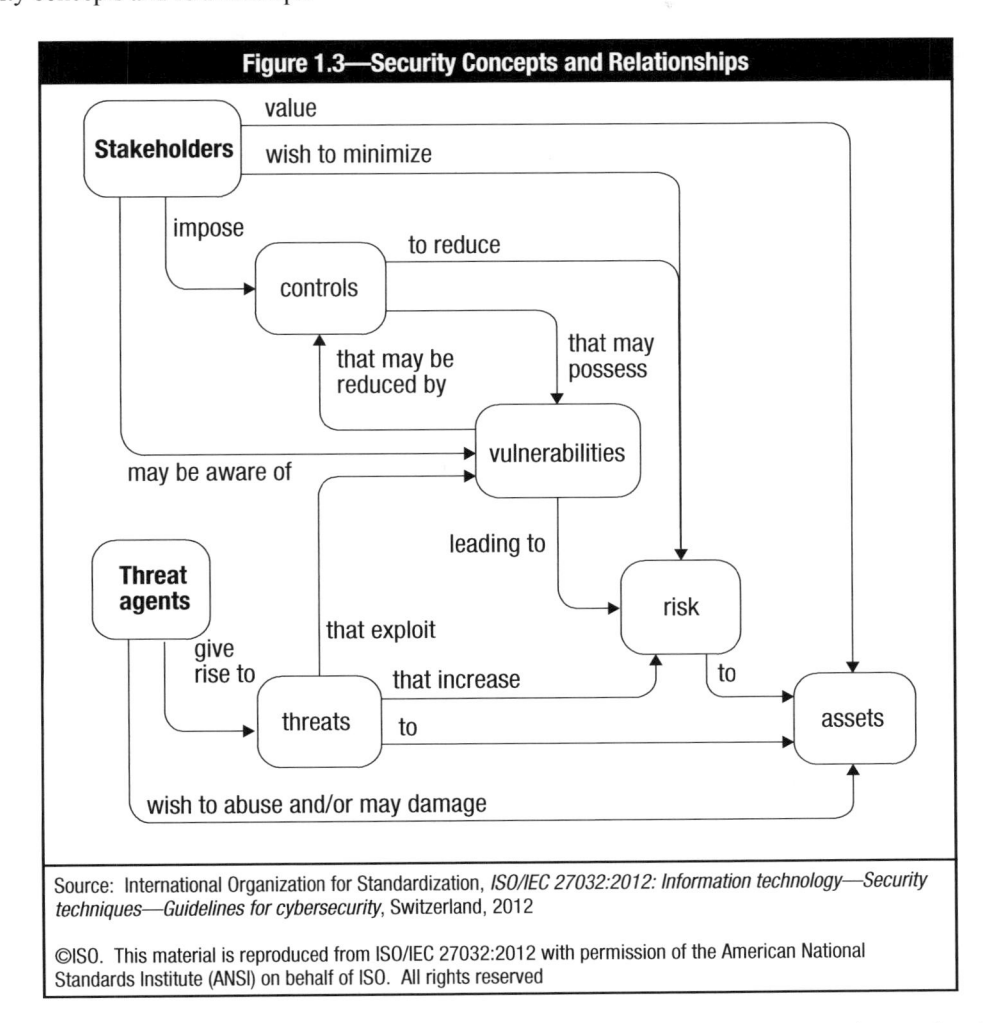

Figure 1.3—Security Concepts and Relationships

- Stakeholders **value** assets
- Stakeholders **wish to minimize** risk
- Stakeholders **impose** controls **to reduce** risk
- Controls **that may be reduced by** vulnerabilities
- Controls **that may possess** vulnerabilities
- Stakeholders **may be aware of** vulnerabilities
- Vulnerabilities **leading to** risk
- Threat agents **give rise to** threats
- Threats **that exploit** vulnerabilities
- Threats **that increase** risk
- Threats **to** assets
- Risk **to** assets
- Threat agents **wish to abuse and/or may damage** assets

Source: International Organization for Standardization, *ISO/IEC 27032:2012: Information technology—Security techniques—Guidelines for cybersecurity*, Switzerland, 2012

©ISO. This material is reproduced from ISO/IEC 27032:2012 with permission of the American National Standards Institute (ANSI) on behalf of ISO. All rights reserved

The responsibility for protecting these assets lies with the stakeholders to whom these assets have value. Therefore, stakeholders should take threats into consideration when assessing risk to these assets. This risk assessment, discussed in section 4.1, will help in the process of control selection. Controls used to protect assets, reduce vulnerabilities and impacts, and/or reduce risk to an acceptable level. They should be monitored and reviewed to ensure that each specific control is adequate to counter the risk it is designed to mitigate. It is important to keep in mind that risk cannot be eliminated, resulting in residual risk. Stakeholders should aim to minimize the level of residual risk. It may also be necessary to use external resources to ensure that controls are functioning properly.

The business environment, in particular, tends to drive risk decisions. For example, a small start-up company may be much more tolerant of risk than a large, well-established corporation. Therefore, it can be helpful to reference broad criteria listed below when evaluating the drivers affecting the security of a specific organization.

With respect to technology, many factors can impact security, such as:
- Level of IT complexity
- Network connectivity (e.g., internal, third-party, public)
- Specialist industry devices/instrumentation
- Platforms, applications and tools used
- On-premise cloud or hybrid systems
- Operational support for security
- User community and capabilities
- New or emerging security tools

When evaluating business plans and the general business environment, consider drivers, such as:
- Nature of business
- Risk tolerance
- Risk appetite
- Security mission, vision and strategy
- Industry alignment and security trends
- Industry-specific compliance requirements and regulations
- Regional regulatory and compliance requirements
  - Country or state of operation
- Mergers, acquisitions and partnerships
  - Consider type, frequency and resulting level of integration
- Outsourcing services or providers

Although business and technology drivers cannot all be predicted with certainty, they should be anticipated reasonably and handled as efficiently as possible. Failure to anticipate key security drivers reflects an inability to effectively react to changing business circumstances, which in turn results in diminished security and missed opportunities for improvement.

THE CYBERSECURITY SKILLS GAP

Cybersecurity is a field that demands skilled professionals who possess the foundational knowledge, education and thought leadership necessary to confront the difficulties that accompany constant technological change. Advanced threat vectors, emerging technologies and a myriad of regulations require cybersecurity professionals to be skilled in technology as well as business and communications.

Cybersecurity addresses both internal and external threats to an organization's digital information assets by focusing on critical electronic data processes, transaction processing, risk analytics and information system security engineering.

There are an estimated 410,000 to 510,000 information security professionals worldwide, and jobs are expected to increase 53 percent by 2018 with over 4.2 million jobs available. However, recent studies and reports suggest that there are simply not enough skilled professionals to fill them.

Section 1: Cybersecurity Introduction and Overview

While the cybersecurity landscape has evolved, the skill set among existing and potential cybersecurity professionals has not kept pace. There is a shortage of skilled professionals in the cybersecurity profession, as seen in **figure 1.4**. According to the ISACA *2017 State of Cybersecurity Study*, 48% of organizations get fewer than 10 applicants for cybersecurity positions, and 64% say fewer than half of their cybersecurity applicants are qualified.[3] Likewise, the European Union's (EU's) digital agenda commissioner believes that the growing cybersecurity skills gap is threatening the EU's competitiveness. Skills gaps are seen in both technical and business aspects of security. This guide provides an overview of these business and technical practices, along with various other methodologies and procedures related to cybersecurity.

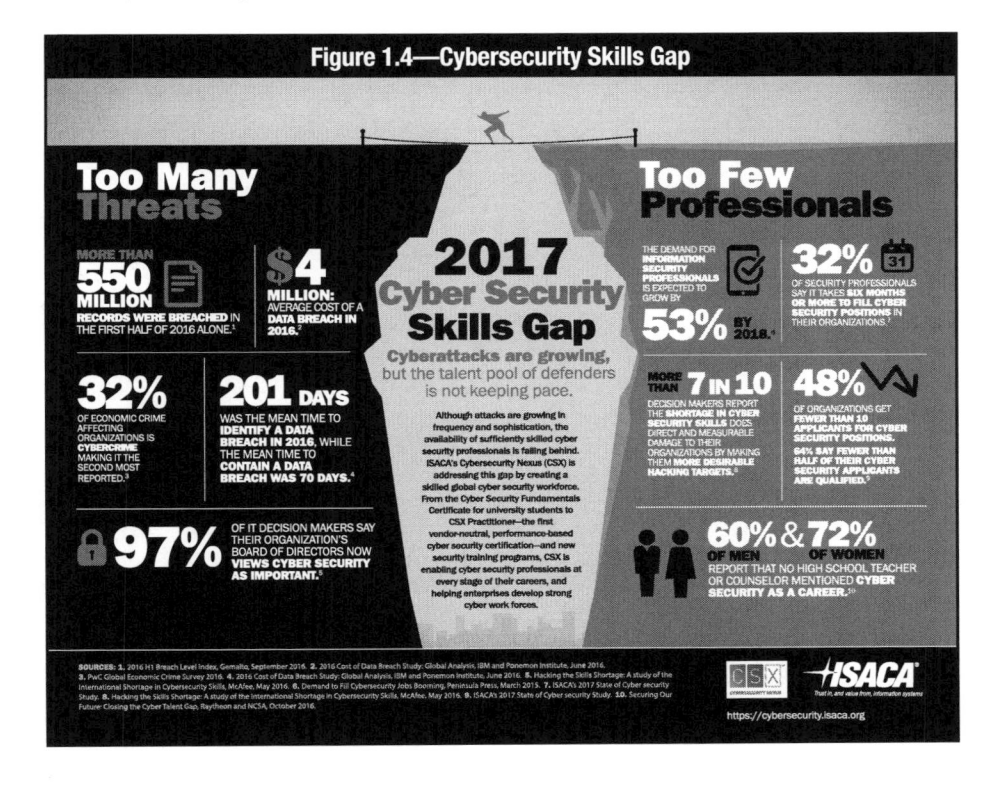

Figure 1.4—Cybersecurity Skills Gap

---

[3] ISACA, *2017 State of Cybersecurity Study, https://cybersecurity.isaca.org*

Page intentionally left blank

TOPIC 2—DIFFERENCE BETWEEN INFORMATION SECURITY AND CYBERSECURITY

Information security deals with information, regardless of its format—it encompasses paper documents, digital and intellectual property in people's minds, and verbal or visual communications. **Cybersecurity**, on the other hand, is concerned with protecting digital assets—everything encompassed within network hardware, software and information that is processed, stored within isolated systems or transported by internetworked information environments. Additionally, concepts such as nation-state-sponsored attacks and APTs belong almost exclusively to cybersecurity. It is helpful to think of cybersecurity as a component of information security.

Therefore, to eliminate confusion, the term cybersecurity will be defined in this guide as the protection of information assets by addressing threats to information processed, stored and transported by internetworked information systems.

PROTECTING DIGITAL ASSETS

In the core of its cybersecurity framework, the National Institute of Standards and Technology (NIST) identifies five key functions necessary for the protection of digital assets. These functions coincide with incident management methodologies and include the following activities:[4]

- **Identify**—Use organizational understanding to minimize risk to systems, assets, data and capabilities.
- **Protect**—Design safeguards to limit the impact of potential events on critical services and infrastructure.
- **Detect**—Implement activities to identify the occurrence of a cybersecurity event.
- **Respond**—Take appropriate action after learning of a security event.
- **Recover**—Plan for resilience and the timely repair of compromised capabilities and services.

---

[4] National Institute of Standards and Technology (NIST), *Framework for Improving Critical Infrastructure Cybersecurity*, USA, 2014, www.nist.gov/sites/default/files/documents/cyberframework/cybersecurity-framework-021214.pdf

Page intentionally left blank

Section 1: Cybersecurity Introduction and Overview

TOPIC 3—CYBERSECURITY OBJECTIVES

Cybersecurity requires stakeholders in the cyberspace area to be active in security, beyond the protection of their own assets. They should be prepared to identify and address emerging risk and challenges to keep assets protected. Cybersecurity works with information security and is beyond merely Internet, network and/or application security. It requires working with all of these components in order to keep the cyberspace useful and trustworthy.

Cybersecurity aims to meet certain objectives in the protection of digital assets. These include the confidentiality, integrity and availability of the assets among other aspects. These are discussed in detail in this topic.

CONFIDENTIALITY, INTEGRITY AND AVAILABILITY

To better understand cybersecurity and the protection of digital assets, it is helpful to consider three key concepts that are used to guide information security policies, as shown in **figure 1.5**. The concepts are:
- Confidentiality
- Integrity
- Availability

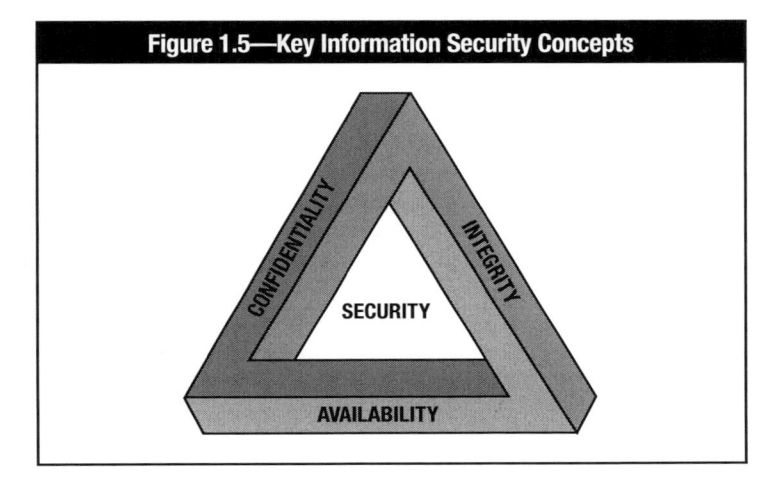

Figure 1.5—Key Information Security Concepts

(Triangle diagram showing CONFIDENTIALITY, INTEGRITY, and AVAILABILITY as the three sides surrounding SECURITY at the center)

Confidentiality is the protection of information from unauthorized access or disclosure. Different types of information require different levels of confidentiality, and the need for confidentiality can change over time. Personal, financial and medical information require a higher degree of confidentiality than the minutes of a staff meeting, for example. Similarly, some companies need to protect information on competitive future products before their release but may need to make the same information public afterward.

Data must be protected from improper disclosure according to its sensitivity and applicable legal requirements. The confidentiality of digital information can be maintained using several different means, including access controls, file permissions and encryption.

Integrity is the protection of information from unauthorized modification. For example, if a bank transfers $10,000 to another financial institution, it is important that the amount does not change to $100,000 during the exchange. The concept of integrity also applies to electronic messaging, files, software and configurations.

Any violation of integrity is significant because it may be the first step in a successful attack against system availability or confidentiality. Contaminated systems and corrupted data must be dealt with immediately to assess the potential for further violation or damage. The integrity of digital assets can be controlled and verified by logging, digital signatures, hashes, encryption and access controls.

Availability ensures the timely and reliable access to and use of information and systems. This includes safeguards to make sure data are not accidentally or maliciously deleted. This is particularly important with a mission-critical system, because any interruptions in its availability can result in a loss of productivity and revenue. Similarly, the loss of data can impact management's ability to make effective decisions and responses. Availability can be protected by the use of redundancy, backups and implementation of business continuity management and planning (see Section 5, Topic 5 for more information).

The impacts, potential consequences and methods of control of confidentiality, integrity and availability are shown in **figure 1.6**.

Figure 1.6—Confidentiality, Integrity and Availability Model and Related Impacts		
Requirement	**Impact and Potential Consequences**	**Methods of Control**
Confidentiality: The protection of information from unauthorized disclosure	Loss of confidentiality can result in the following consequences: • Disclosure of information protected by privacy laws • Loss of public confidence • Loss of competitive advantage • Legal action against the enterprise • Interference with national security • Loss of compliance	Confidentiality can be preserved using the following methods: • Access controls • File permissions • Encryption
Integrity: The accuracy and completeness of information in accordance with business values and expectations	Loss of integrity can result in the following consequences: • Inaccuracy • Erroneous decisions • Fraud • Failure of hardware • Loss of compliance	Integrity can be preserved using the following methods: • Access controls • Logging • Digital signatures • Hashes • Backups • Encryption
Availability: The ability to access information and resources required by the business process	Loss of availability can result in the following consequences: • Loss of functionality and operational effectiveness • Loss of productive time • Fines from regulators or a lawsuit • Interference with enterprise's objectives • Loss of compliance	Availability can be preserved using the following methods: • Redundancy of network, system, data • Highly available system architectures • Data replication • Backups • Access controls • A well-designed disaster recovery plan or business continuity plan

NONREPUDIATION

Nonrepudiation is an important consideration in cybersecurity. It refers to the concept of ensuring that a message or other piece of information is genuine. When information is sent, it is important to verify that it is coming from the source that it says it is coming from. Nonrepudiation provides a means so that the person who sends or receives information cannot deny that they sent or received the information. It is implemented through digital signatures and transactional logs.

TOPIC 4—CYBERSECURITY GOVERNANCE

GOVERNANCE, RISK MANAGEMENT AND COMPLIANCE

The structure and governance of every organization is different and varies based on the type of organization. Each organization has its own mission (business), size, industry, culture and legal regulations. However, all organizations have a responsibility and duty to protect their assets and operations, including their IT infrastructure and information. At the highest level, this is generally referred to as governance, risk management and compliance (GRC). Some entities implement these three areas in an integrated manner, while others may have less comprehensive approaches. Regardless of the actual implementation, every organization needs a plan to manage these three elements.

Governance is the responsibility of the board of directors and senior management of the organization. A governance program has several goals:
- Provide strategic direction
- Ensure that objectives are achieved
- Ascertain whether risk is being managed appropriately
- Verify that the organization's resources are being used responsibly

Risk management is the coordination of activities that direct and control an enterprise with regard to risk. Risk management requires the development and implementation of internal controls to manage and mitigate risk throughout the organization, including financial, operational, reputational, and investment risk, physical risk and cyberrisk.

Compliance is the act of adhering to, and the ability to demonstrate adherence to, mandated requirements defined by laws and regulations. It also includes voluntary requirements resulting from contractual obligations and internal policies.

Cybersecurity is the responsibility of the entire organization at every level. The next section outlines some of the specific roles in managing cyberrisk within most organizations.

ROLE OF THE CYBERSECURITY PROFESSIONAL

The cybersecurity professional's duties include analysis of policy, trends and intelligence. Using problem solving and detection skills, they strive to better understand how an adversary may think or behave. The inherent complexity of their work requires the cybersecurity workforce to possess not only a wide array of technical IT skills, but also advanced analytical capabilities. A cybersecurity professional may be a practitioner and/or part of senior management.

INFORMATION SECURITY ROLES

Because cybersecurity is part of information security, there is occasional overlap between the terms and how they are applied to management structures and titles. For the purposes of this discussion, assume that the term information security encompasses cybersecurity roles and functions.

Board of Directors

Cybersecurity governance requires strategic direction and inputs. It depends on commitment, resources and responsibility for cybersecurity management, and it requires a means for the board to determine whether its intent has been met. Effective governance can be accomplished only by senior management involvement in approving policy and by appropriate monitoring and metrics coupled with reporting and trend analysis.

Members of the board need to be aware of the organization's information assets and their criticality to ongoing business operations. The board should periodically be provided with the high-level results of comprehensive risk assessments and business impact analyses (BIAs), which identify how quickly essential business unit and processes have to return to full operation following a disaster event. As a result of these activities, board members should identify the key assets they want protected and verify that protection levels and priorities are appropriate to a standard of due care.

Section 1: Cybersecurity Introduction and Overview

The tone at the top must be conducive to effective security governance. It is up to senior management to set a positive example in this regard, as lower-level personnel are much more likely to abide by security measures when they see their superiors respecting the same measures as well. Executive management's endorsement of security requirements ensures that security expectations are met at all levels of the enterprise. Penalties for noncompliance must be defined, communicated and enforced from the board level down.

Beyond these requirements, the board has an ongoing obligation to provide oversight for activities related to cybersecurity. Senior management has an ethical responsibility, and in many cases a legal obligation, to exercise due care in protection of the organization's key assets, including its confidential and critical information. Therefore, their involvement and oversight is required.

Executive Management

An organization's executive management team is responsible for ensuring that needed organizational functions, resources and supporting infrastructure are available and properly utilized to fulfill the directives of the board, regulatory compliance and other demands.

Generally, executive management looks to the chief information security officer (CISO) or other senior cybersecurity manager to define the information security program and its subsequent management. Often, the cybersecurity manager is also expected to provide education and guidance to the executive management team. As opposed to being the decision maker, the manager's role in this situation is often constrained to the presentation of options supported by key, actionable information. In other words, the cybersecurity manager acts as an advisor.

Executive management sets the tone for cybersecurity management within the organization. The level of visible involvement and the inclusion of information risk management in key business activities and decisions indicate to other managers the level of importance that they are also expected to apply to risk management for activities within their organizations.

Senior Information Security Management

The exact title for the individual who oversees information security and cybersecurity varies from organization to organization. One of the most common titles is CISO, but some organizations prefer chief security officer (CSO) to denote responsibility for all security matters, both physical and digital. Likewise, the responsibilities and authority of information security managers vary dramatically between organizations.

Generally, the cybersecurity manager will be responsible for:
- Developing the security strategy
- Overseeing the security program and initiatives
- Coordinating with business process owners for ongoing alignment
- Ensuring that risk and business impact assessments are conducted
- Developing risk mitigation strategies
- Enforcing policy and regulatory compliance
- Monitoring the utilization and effectiveness of security resources
- Developing and implementing monitoring and metrics
- Directing and monitoring security activities
- Managing cybersecurity incidents and their remediation, as well as incorporating lessons learned

Cybersecurity Practitioners

In most organizations, cybersecurity is managed by a team of subject matter experts and cybersecurity practitioners, including security architects, administrators, digital forensics, incident handlers, vulnerability researchers, and network security specialists. Together they design, implement and manage processes and technical controls and respond to events and incidents.

These practitioners work within the direction, policies, guidelines, mandates and regulations set by the board of directors, executives and cybersecurity management. Cybersecurity roles are shown in **figure 1.7**.

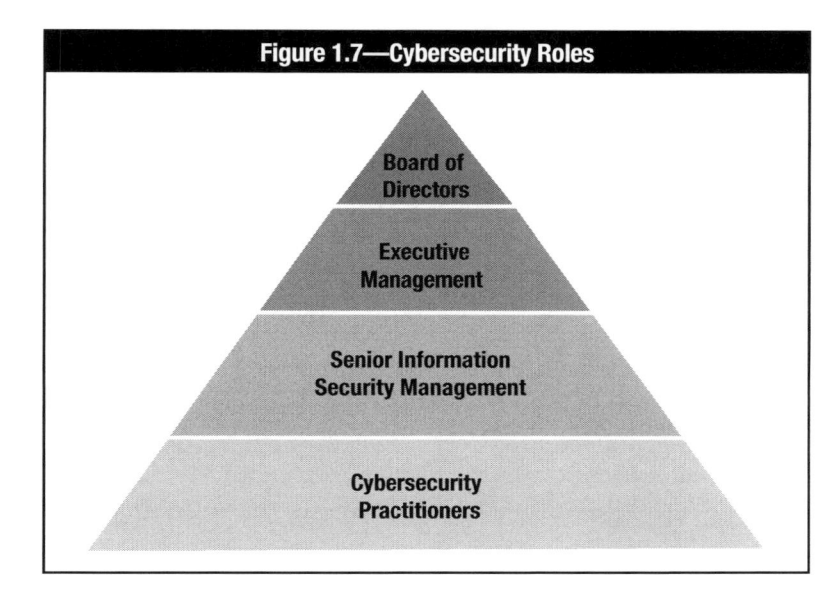

Figure 1.7—Cybersecurity Roles

- Board of Directors
- Executive Management
- Senior Information Security Management
- Cybersecurity Practitioners

Page intentionally left blank

TOPIC 5—CYBERSECURITY DOMAINS

There are five cybersecurity domains. Each domain is covered in detail its own section in this guide. This topic provides an overview of each domain. The five cybersecurity domains are:
- Cybersecurity Concepts
- Security Architecture Principles
- Security of Networks, Systems, Applications and Data
- Incident Response
- Security Implications and Adoption of Evolving Technology

CYBERSECURITY CONCEPTS

This domain provides discussion of critical concepts such as:
- Basic risk management
- Common attack vectors and threat agents
- Patterns and types of attacks
- Types of security policies and procedures
- Cybersecurity control processes

All of these concepts are addressed in regard to how they influence security policies and procedures relating to cybersecurity threats. Each topic considers various approaches with a focus on security best practices.

SECURITY ARCHITECTURE PRINCIPLES

This domain provides information that helps security professionals identify and apply the principles of security architecture. It discusses a variety of topics, including:
- Common security architectures and frameworks
- Perimeter security concepts
- System topology and perimeter concepts
- Firewalls and encryption
- Isolation and segmentation
- Methods for monitoring, detection and logging

These topics are presented with a focus on best security practice. Various types of security architectures are discussed to illustrate the importance of layering controls to achieve defense in depth.

SECURITY OF NETWORKS, SYSTEMS, APPLICATIONS AND DATA

This domain addresses basic system hardening techniques and security measures, including:
- Process controls
  - Risk assessments
  - Vulnerability management
  - Penetration testing
- Best practices for securing networks, systems, applications and data
  - System and application security threats and vulnerabilities
  - Effective controls for managing vulnerabilities

These discussions aim to help cybersecurity professionals assess their risk tolerance and respond appropriately to vulnerabilities.

INCIDENT RESPONSE

This domain articulates the critical distinction between an event and an incident. More important, it outlines the steps necessary when responding to a cybersecurity incident. It covers the following topics:
- Incident categories
- Disaster recovery and business continuity plans
- Steps of incident response
- Forensics and preservation of evidence

These discussions aim to provide entry-level professionals with the level of knowledge necessary to respond to cybersecurity incidents competently.

SECURITY IMPLICATIONS AND ADOPTION OF EVOLVING TECHNOLOGY

This domain outlines the current threat landscape, including a discussion of vulnerabilities associated with the following emerging technologies:
- Mobile devices (bring your own device [BYOD], Internet of Things [IoT])
- Cloud computing and storage
- Digital collaboration (social media)

Although the current threat landscape continues to evolve, this section highlights the recent developments most likely to impact cybersecurity professionals. For example, it discusses the implications of BYOD environments and addresses the risk introduced by mobile and web applications. There is also an extended discussion of APTs and their most frequent targets.

SECTION 1—KNOWLEDGE CHECK

1. Three common controls used to protect the availability of information are:
   A. redundancy, backups and access controls.
   B. encryption, file permissions and access controls.
   C. access controls, logging and digital signatures.
   D. hashes, logging and backups.

2. Select all that apply. Governance has several goals, including:
   A. providing strategic direction.
   B. ensuring that objectives are achieved.
   C. verifying that organizational resources are being used appropriately.
   D. directing and monitoring security activities.
   E. ascertaining whether risk is being managed properly.

3. Choose three. According to the NIST cybersecurity framework, which of the following are considered key functions necessary for the protection of digital assets?
   A. Encrypt
   B. Protect
   C. Investigate
   D. Recover
   E. Identify

4. Which of the following is the best definition for cybersecurity?
   A. The process by which an organization manages cybersecurity risk to an acceptable level
   B. The protection of information from unauthorized access or disclosure
   C. The protection of paper documents, digital and intellectual property, and verbal or visual communications
   D. Protecting information assets by addressing threats to information that is processed, stored or transported by internetworked information systems

5. Which of the following cybersecurity roles is charged with the duty of managing incidents and remediation?
   A. Board of directors
   B. Executive committee
   C. Security management
   D. Cybersecurity practitioners

See answers in Appendix C.

Page intentionally left blank

Section 2: Cybersecurity Concepts

Topics covered in this section include:
1. Risk
2. Common attack types and vectors
3. Policies
4. Cybersecurity controls

Page intentionally left blank

Section 2: Cybersecurity Concepts

TOPIC 1—RISK

The core duty of cybersecurity is to identify, mitigate and manage cyberrisk to an organization's digital assets. Cyberrisk is that portion of overall risk management that solely focuses on risk that manifests in the cyber (Interconnected Information Environments) domain. While most people have an inherent understanding of risk in their day-to-day lives, it is important to understand risk in the context of cybersecurity, which means knowing how to determine, measure and reduce risk effectively.

Assessing risk is one of the most critical functions of a cybersecurity organization. Effective policies, security implementations, resource allocation and incident response preparedness are all dependent on understanding the risk and threats an organization faces. Using a risk-based approach to cybersecurity allows more informed decision-making to protect the organization and to apply limited budgets and resources effectively. If controls are not implemented based on awareness of actual risk, then valuable organizational assets will not be adequately protected while other assets will be wastefully overprotected.[5]

Too often, cybersecurity controls are implemented with little or no assessment of risk. ISACA's worldwide survey of IT management, auditors and security managers showed that over 80 percent of companies believe "information security risks are either not known or are only partially assessed" and that "IT risk illiteracy and lack of awareness" are major challenges in managing risk.[6] Therefore, understanding risk and risk assessments are critical requirements for any security practitioner.

KEY TERMS AND DEFINITIONS

A visual summary of the key terms is presented in International Organization for Standardization (ISO)/International Electrotechnical Commission (IEC) 27032 (**figure 2.1**).

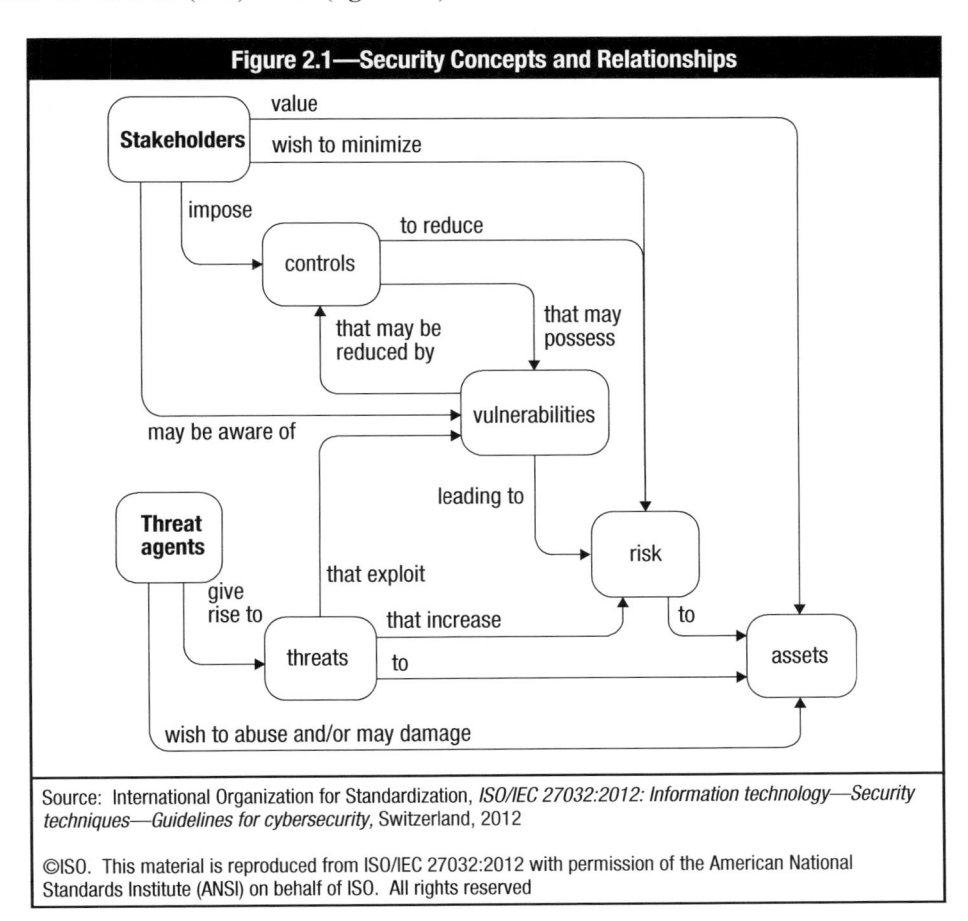

Figure 2.1—Security Concepts and Relationships

- Stakeholders *value* assets
- Stakeholders *wish to minimize* risk
- Stakeholders *impose* controls
- controls *to reduce* risk
- controls *that may be reduced by* vulnerabilities
- vulnerabilities *that may possess* (assets)
- vulnerabilities *leading to* risk
- Stakeholders *may be aware of* vulnerabilities
- Threat agents *give rise to* threats
- Threat agents *wish to abuse and/or may damage* assets
- threats *that exploit* vulnerabilities
- threats *that increase* risk
- threats *to* assets
- risk *to* assets

Source: International Organization for Standardization, *ISO/IEC 27032:2012: Information technology—Security techniques—Guidelines for cybersecurity,* Switzerland, 2012

©ISO. This material is reproduced from ISO/IEC 27032:2012 with permission of the American National Standards Institute (ANSI) on behalf of ISO. All rights reserved

---

[5] Anderson, Kent, "A Business Model for Information Security," *ISACA® Journal*, Vol. 3, 2008
[6] ISACA, "Top Business/Security Issues Survey Results," USA, 2011

Section 2: Cybersecurity Concepts

There are many potential definitions of risk—some general and others more technical. Additionally, it is important to distinguish between a risk and a threat. Although many people use the words threat and risk synonymously, they have two very different meanings. As with any key concept, there is some variation in definition from one organization to another. For the purposes of this guide, we will define terms as follows:

- **Risk**—The combination of the probability of an event and its consequence (ISO/IEC 73). Risk is mitigated through the use of controls or safeguards.
- **Threat**—Anything (e.g., object, substance, human) that is capable of acting against an asset in a manner that can result in harm. ISO/IEC 13335 defines a threat broadly as a potential cause of an unwanted incident. Some organizations make a further distinction between a threat source and a threat event, classifying a threat source as the actual process or agent attempting to cause harm, and a threat event as the result or outcome of a threat agent's malicious activity.
- **Asset**—Something of either tangible or intangible value that is worth protecting, including people, information, infrastructure, finances and reputation
- **Vulnerability**—A weakness in the design, implementation, operation or internal control of a process that could expose the system to adverse threats from threat events. Although much of cybersecurity is focused on the design, implementation and management of controls to mitigate risk, it is critical for security practitioners to understand that risk can never be eliminated. Beyond the general definition of risk provided above, there are other, more specific types of risk that apply to cybersecurity.
- **Inherent risk**—The risk level or exposure without taking into account the actions that management has taken or might take (e.g., implementing controls)
- **Residual risk**—Even after safeguards are in place, there will always be residual risk, defined as the remaining risk after management has implemented a risk response.

Figure 2.2 illustrates one example of how many of these key terms come into play when framing an approach to risk management.

Figure 2.2—Framing Risk Management

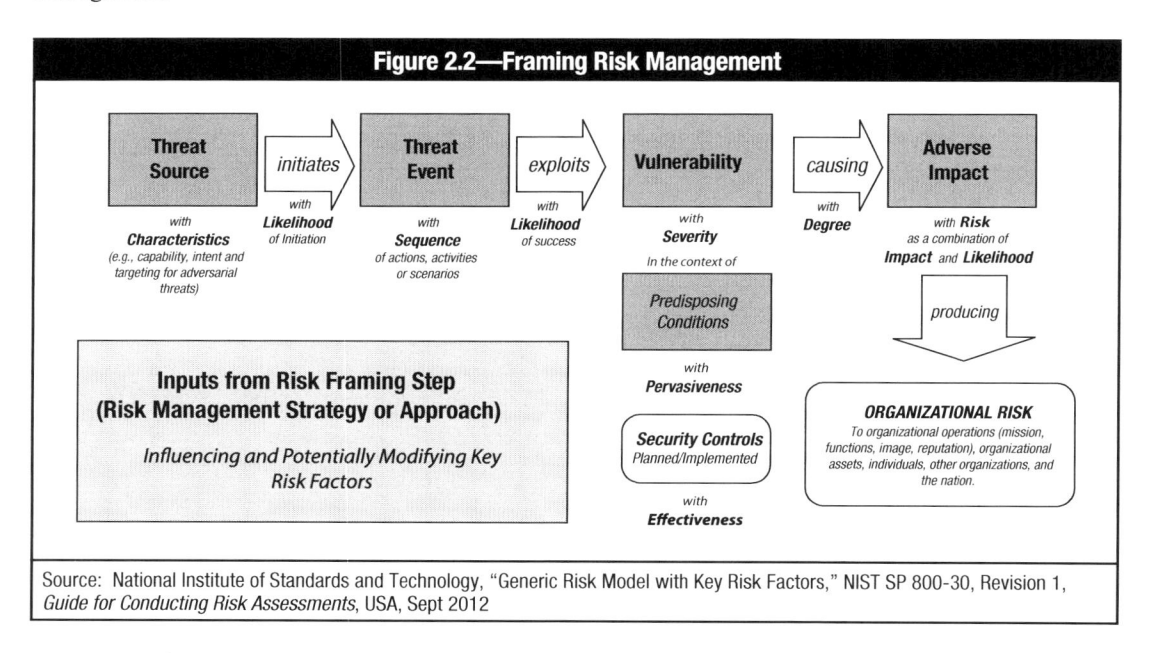

Threat Source — initiates → Threat Event — exploits → Vulnerability — causing → Adverse Impact

- Threat Source: with Characteristics (e.g., capability, intent and targeting for adversarial threats); with Likelihood of Initiation
- Threat Event: with Sequence of actions, activities or scenarios; with Likelihood of success
- Vulnerability: with Severity, In the context of Predisposing Conditions (with Pervasiveness) and Security Controls Planned/Implemented (with Effectiveness)
- Adverse Impact: with Degree; with Risk as a combination of Impact and Likelihood

producing → **ORGANIZATIONAL RISK**
To organizational operations (mission, functions, image, reputation), organizational assets, individuals, other organizations, and the nation.

Inputs from Risk Framing Step (Risk Management Strategy or Approach)
Influencing and Potentially Modifying Key Risk Factors

Source: National Institute of Standards and Technology, "Generic Risk Model with Key Risk Factors," NIST SP 800-30, Revision 1, *Guide for Conducting Risk Assessments*, USA, Sept 2012

Risk Identification and Classification Standards and Frameworks[7]

Several good sources for risk identification and classification standards and frameworks are available to the cybersecurity professional. The following list is not comprehensive, and many more standards are available. However, this list may allow the cybersecurity professional to consider a framework or standard that would be suitable for use in his/her organization. Many countries and industries have specific standards that must be used by organizations operating in their jurisdiction. The use of a recognized standard may provide credibility and

---

[7] ISACA, *CRISC Review Manual 6th Edition*, USA, 2015

completeness for the risk assessment and management program of the organization and help ensure that the risk management program is comprehensive and thorough.

ISO 31000:2009 Risk Management—Principles and Guidelines
ISO 31000:2009 states:

> This international standard recommends that organizations develop, implement and continuously improve a framework whose purpose is to integrate the process for managing risk into the organization's overall governance, strategy and planning, management, reporting polices, values and culture.
>
> Although the practice of risk management has been developed over time and within many sectors in order to meet diverse needs, the adoption of consistent processes within a comprehensive framework can help to ensure that risk is managed effectively, efficiently and coherently across an organization. The generic approach described in this standard provides the principles and guidelines for managing any form of risk in a systematic, transparent and credible manner and within any scope and context.[8]

COBIT® 5 for Risk
COBIT 5 for Risk is described as follows:

> COBIT 5 provides a comprehensive framework that assists enterprises in achieving their objectives for the governance and management of enterprise information technology (IT). Simply stated, COBIT 5 helps enterprises to create optimal value from IT by maintaining a balance between realising benefits and optimising risk levels and resource use. COBIT 5 enables IT to be governed and managed in a holistic manner for the entire enterprise, taking into account the full end-to-end business and IT functional areas of responsibility and considering the IT-related interests of internal and external stakeholders.
>
> COBIT 5 for Risk ... builds on the COBIT 5 framework by focusing on risk and providing more detailed and practical guidance for risk professionals and other interested parties at all levels of the enterprise.[9]

IEC 31010:2009 Risk Management—Risk Assessment Techniques
IEC 31010:2009 states:

> Organizations of all types and sizes face a range of risks that may affect the achievement of their objectives.
>
> These objectives may relate to a range of the organization's activities, from strategic initiatives to its operations, processes and projects, and be reflected in terms of societal environmental, technological, safety and security outcomes, commercial, financial and economic measures, as well as social, cultural, political and reputation impacts.
>
> All activities of an organization involve risks that should be managed. The risk management process aids decision making by taking account of uncertainty and the possibility of future events or circumstances (intended or unintended) and their effects on agreed objectives.[10]

ISO/IEC 27001:2013 Information Technology—Security Techniques—Information Security Management Systems—Requirements
ISO 27001:2013 states:

> The organization shall define and apply an information security risk assessment process that: c) identifies the information security risks:
> 1) apply the information security risk assessment process to identify risks associated with the loss of confidentiality, integrity and availability for information within the scope of the information security management system; and
> 2) identify risk owners.[11]

---

[8] ISO; *ISO 31000:2009 Risk Management—Principles and Guidelines*, Switzerland, 2009
[9] ISACA, *COBIT 5 for Risk*, USA, 2013
[10] ISO; *IEC 31010:2009 Risk Management—Risk Assessment Techniques*, Switzerland, 2009
[11] ISO; *ISO/IEC 27001:2013 Information Technology—Security Techniques—Information Security Management Systems—Requirements*, Switzerland, 2013

ISO/IEC 27005:2011 Information Technology—Security Techniques—Information Security Risk Management
ISO/IEC 27005 states:

This international standard provides guidelines for information security risk management in an organization, supporting in particular the requirements of an information security management system (ISMS) according to ISO/IEC 27001. However, this standard does not provide any specific methodology for information security risk management. It is up to the organization to define their approach to risk management, depending for example on the scope of the ISMS, context of risk management, or industry sector. A number of existing methodologies can be used under the framework described in this International standard to implement the requirements of an ISMS.[12]

NIST Special Publications
NIST has a wide range of special publications available at csrc.nist.gov. Some of the publications related to IT risk are discussed in the following sections.

NIST Special Publication 800-30 Revision 1: Guide for Conducting Risk Assessments
NIST Special Publication 800-30 Revision 1 describes risk assessment in the following manner: Risk assessments are a key part of effective risk management and facilitate decision making at all three tiers in the risk management hierarchy including the organization level, mission/business process level, and information system level.

Because risk management is ongoing, risk assessments are conducted throughout the system development life cycle, from pre-system acquisition (i.e., material solution analysis and technology development), through system acquisition (i.e., engineering/manufacturing development and production/deployment), and on into sustainment (i.e., operations/support).[13]

NIST Special Publication 800-39: Managing Information Security Risk
NIST Special Publication 800-39 states:

The purpose of Special Publication 800-39 is to provide guidance for an integrated, organization-wide program for managing information security risk to organizational operations (i.e., mission, functions, image, and reputation), organizational assets, individuals, other organizations, and the Nation resulting from the operation and use of federal information systems. Special Publication 800-39 provides a structured, yet flexible approach for managing risk that is intentionally broad-based, with the specific details of assessing, responding to, and monitoring risk on an ongoing basis provided by other supporting NIST security standards and guidelines.[14]

Risk Identification (Risk Scenarios)[15]
A risk scenario is a description of a possible event whose occurrence will have an uncertain impact on the achievement of the enterprise's objectives, which may be positive or negative. The development of risk scenarios provides a way of conceptualizing risk that can aid in the process of risk identification. Scenarios are also used to document risk in relation to business objectives or operations impacted by events, making them useful as the basis for quantitative risk assessment. Each identified risk should be included in one or more scenarios, and each scenario should be based on an identified risk.

The development of risk scenarios is based on describing a potential risk event and documenting the factors and areas that may be affected by the risk event. Each scenario should be related to a business objective or impact. Risk events may include system failure, loss of key personnel, theft, network outages, power failures, or any other situation that could affect business operations and mission. The key to developing effective scenarios is to focus on real and relevant potential risk events.

---
[12] ISO/IEC; *ISO/IEC 27005:2011 Information Technology—Security Techniques—Information Security Risk Management*, Switzerland, 2011.
[13] NIST; *NIST Special Publication 800-30 Revision 1: Guide for Conducting Risk Assessments*, USA, 2012
[14] NIST; *NIST Special Publication 800-39: Managing Information Security Risk*, USA, 2011
[15] ISACA, *COBIT 5 for Risk*, USA, 2013

The development of risk scenarios purely from imagination is an art that often requires creativity, thought, consultation and questioning. Incidents that have occurred previously may be used as the basis of risk scenarios with far less effort put into their development. Risk scenarios based on past events should be fully explored to ensure that similar situations do not recur in ways that might have been avoided. Risk scenarios can be developed from a top-down perspective driven by business goals or from a bottom-up perspective originating from several inputs, as shown in **figure 2.3**.

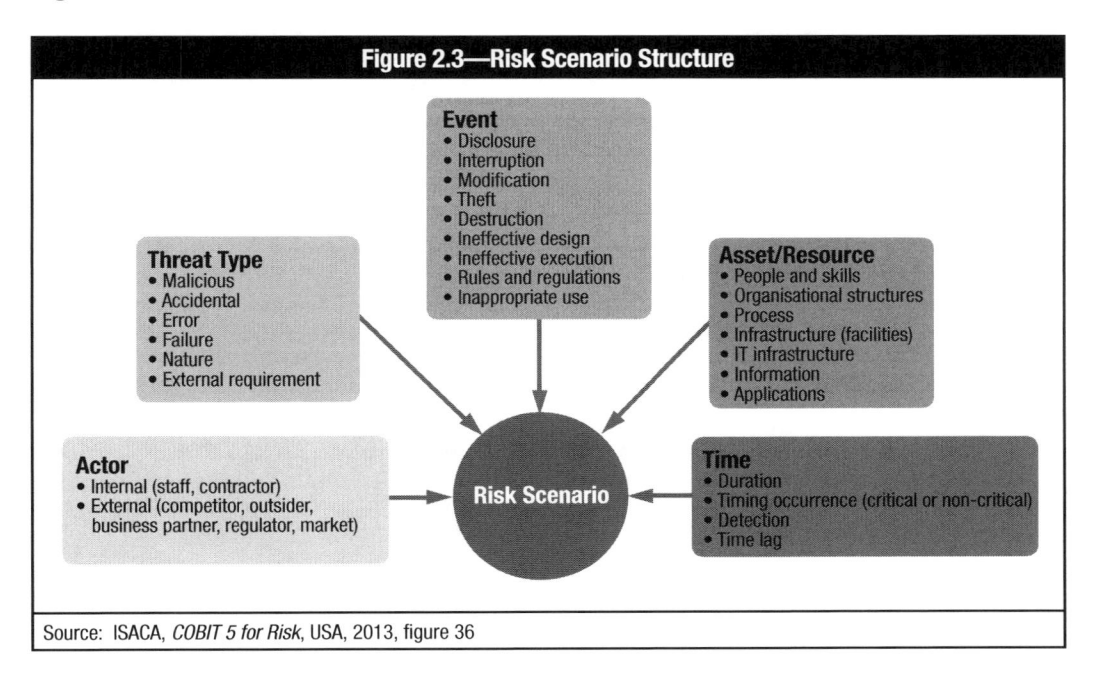

Figure 2.3—Risk Scenario Structure

Threat Type
- Malicious
- Accidental
- Error
- Failure
- Nature
- External requirement

Event
- Disclosure
- Interruption
- Modification
- Theft
- Destruction
- Ineffective design
- Ineffective execution
- Rules and regulations
- Inappropriate use

Asset/Resource
- People and skills
- Organisational structures
- Process
- Infrastructure (facilities)
- IT infrastructure
- Information
- Applications

Actor
- Internal (staff, contractor)
- External (competitor, outsider, business partner, regulator, market)

Time
- Duration
- Timing occurrence (critical or non-critical)
- Detection
- Time lag

Source: ISACA, *COBIT 5 for Risk*, USA, 2013, figure 36

Top-down Approach
A top-down approach to scenario development is based on understanding business goals and how a risk event could affect the achievement of those goals. Under this model, the risk practitioner looks for the outcome of events that may hamper business goals identified by senior management. Various scenarios are developed that allow the organization to examine the relationship between the risk event and the business goals, so that the impact of the risk event can be measured. By directly relating a risk scenario to the business, senior managers can be educated and involved in how to understand and measure risk.

The top-down approach is suited to general risk management of the company, because it looks at both IT- and non-IT-related events. A benefit of this approach is that because it is more general, it is easier to achieve management buy-in even if management usually is not interested in IT. The top-down approach also deals with the goals that senior managers have already identified as important to them.

Bottom-up Approach
The bottom-up approach to developing risk scenarios is based on describing risk events that are specific to cybersecurity-related situations, typically hypothetical situations envisioned by the people performing the job functions in specific processes. The cybersecurity professional and assessment team start with one or more generic risk scenarios then refine them to meet their individual organizational needs, including building complex scenarios to account for coinciding events.

Bottom-up scenario development can be a good way to identify scenarios that are highly dependent on the specific technical workings of a process or system, which may not be apparent to anyone who is not intimately involved with that work but could have substantial consequences for the organization. One downside of bottom-up scenario development is that it may be more difficult to maintain management interest in highly specialized technical scenarios.

Section 2: Cybersecurity Concepts

LIKELIHOOD AND IMPACT[16]

Likelihood (also called probability) is the measure of frequency of which an event may occur, which depends on whether there is a potential source for the event (threat) and the extent to which the particular type of event can affect its target (vulnerability), taking into account any controls or countermeasures that the organization has put in place to reduce its vulnerability. In the context of risk identification, likelihood is used to calculate the risk that an organization faces based on the number of events that may occur within a given time period (often an annual basis).

The risk faced by organizations consists of some combination of known and unknown threats, directed against systems that have some combination of known and unknown vulnerabilities. A vulnerability assessment that identifies no vulnerabilities is not equal to the system being invulnerable in the absolute sense. It means only that the types of vulnerabilities that the assessment was intended to detect were not detected. Failure to detect a vulnerability may be the result of its absence, or it may be a false negative arising from misconfiguration of a tool or improper performance of a manual review. In the case of a zero-knowledge penetration test that fails to identify opportunities to exploit a system, it may be that the team was unlucky or lacked imagination, which may not be true of an outside attacker. Even if no known vulnerabilities exist within the system, the system may still remain vulnerable to unknown vulnerabilities, more of which are discovered every day (and some of these are stored or sold for future use as "zero-day" exploits). The likelihood of an attack is often a component of external factors, such as the motivation of the attacker, as shown in **figure 2.4**.

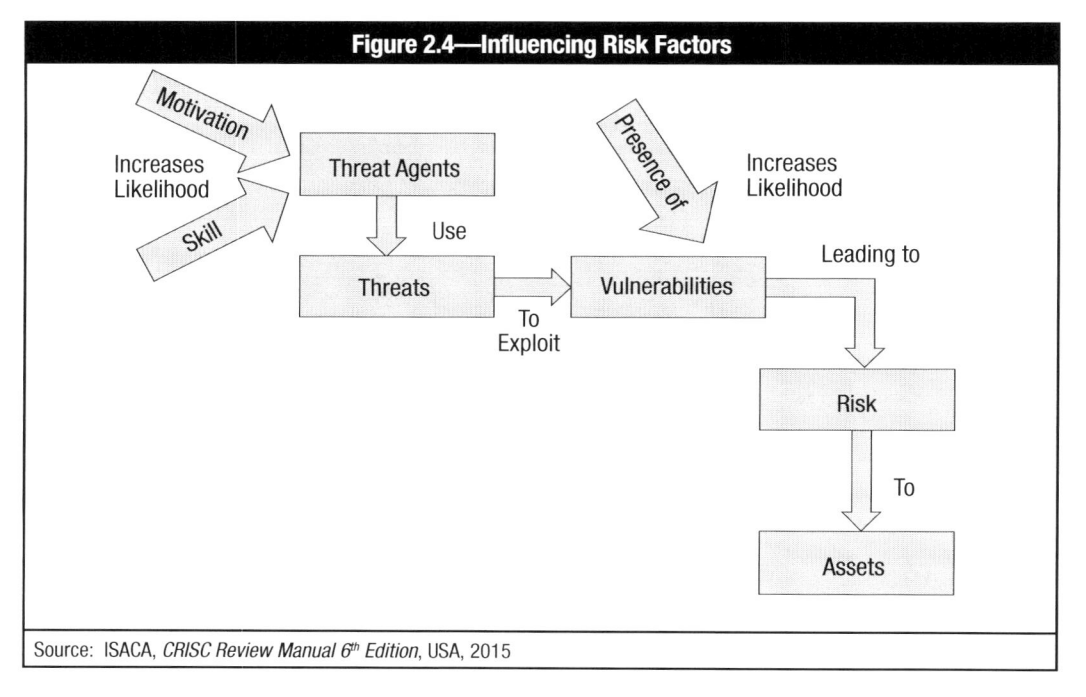

Figure 2.4—Influencing Risk Factors

Motivation and Skill Increase Likelihood → Threat Agents → Use → Threats → To Exploit → Vulnerabilities (Presence of Increases Likelihood) → Leading to → Risk → To → Assets

Source: ISACA, *CRISC Review Manual 6th Edition*, USA, 2015

Given the combination of unknown threat and unknown vulnerability, it is difficult for the cybersecurity professional to provide a comprehensive estimate of the likelihood of a successful attack. Vulnerability assessments and penetration tests provide the cybersecurity practitioner with valuable information on which to partially estimate this likelihood, because:
- Although the presence of a vulnerability does not guarantee a corresponding threat, the all-hours nature of information systems and the rapid speed of processing makes it much more likely that an information system will come under an assortment of attacks in a short time than would be true of a physical system.
- A vulnerability known to an assessment tool is also knowable to threat agents, all but the most elite of whom tend to build their attacks to target common vulnerabilities.
- The presence of one or more known vulnerabilities—unless these have been previously identified and the risk is accepted by the organization for good reason—suggests a weakness in the overall security program.

---

[16] ISACA, *CRISC Review Manual 6th Edition*, USA, 2015

When assessing a threat, cybersecurity professionals often analyze the threat's likelihood and impact in order to rank and prioritize it among other existing threats.

In some cases where clear, statistically sound data are available, likelihood can be a matter of mathematical probability. This is true with situations such as weather events or natural disasters. However, sometimes accurate data are simply not available, as is often the case when analyzing human threat agents in cybersecurity environments. Some factors create situations where the likelihood of certain threats is more or less prevalent for a given organization. For example, a connection to the Internet will predispose a system to port scanning. Typically, qualitative rankings such as "High, Medium, Low" or "Certain, Very Likely, Unlikely, Impossible" can be used to rank and prioritize threats stemming from human activity. When using qualitative rankings, however, the most important step is to rigorously define the meaning of each category and use definitions consistently throughout the assessment process.

For each identified threat, the impact or magnitude of harm expected to result should also be determined. The impact of a threat can take many forms, but it often has an operational consequence of some sort, whether financial, reputational or legal. Impacts can be described either qualitatively or quantitatively, but as with likelihoods, qualitative rankings are most often used in cybersecurity risk assessment. Likewise, each ranking should be well-defined and consistently used. In cybersecurity, impacts are also evaluated in terms of confidentiality, integrity and availability.

The cybersecurity professional should ensure that senior management does not develop a false sense of cybersecurity as a result of vulnerability assessments and penetration tests that fail to find vulnerabilities, but both forms of testing do provide insight into the organization and its security posture.

APPROACHES TO RISK

A number of methodologies are available to measure risk. Different industries and professions have adopted various tactics based upon the following criteria:
- Risk tolerance
- Size and scope of the environment in question
- Amount of data available

It is particularly important to understand an organization's risk tolerance when considering how to measure risk. For example, a general approach to measuring risk is typically sufficient for risk-tolerant organizations such as academic institutions or small businesses. However, more rigorous and in-depth risk assessment is required for entities with a low tolerance for risk. This is especially relevant for any heavily regulated entity, like a financial institution or an airline reservation system, where any down time would have a significant operational impact

APPROACHES TO CYBERSECURITY RISK

There are three different approaches to implementing cybersecurity. Each approach is described briefly below:
- *Ad hoc*—An *ad hoc* approach simply implements security with no particular rationale or criteria. Ad hoc implementations may be driven by vendor marketing, or they may reflect insufficient subject matter expertise, knowledge or training when designing and implementing safeguards.
- **Compliance-based**—Also known as standards-based security, this approach relies on regulations or standards to determine security implementations. Controls are implemented regardless of their applicability or necessity, which often leads to a "checklist" attitude toward security.
- **Risk-based**—Risk-based security relies on identifying the unique risk a particular organization faces and designing and implementing security controls to address that risk above and beyond the entity's risk tolerance and business needs. The risk-based approach is usually scenario-based.

In reality, most organizations with mature security programs use a combination of risk-based and compliance-based approaches. In fact, most standards or regulations such as ISO 27001, the Payment Card Industry Data Security Standard (PCI DSS), Sarbanes–Oxley Act (SOX) or the US Health Insurance Portability and Accountability Act (HIPAA) require risk assessments to drive the particular implementation of the required controls.

THIRD-PARTY RISK

Cybersecurity can be more difficult to control when third parties are involved, especially when different entities have different security cultures and risk tolerances. No organization exists in a vacuum, and information must be shared with other individuals or organizations, often referred to as third parties. It is important to understand third-party risk, such as information sharing and network access, as it relates to cybersecurity.

Outsourcing is common, both onshore and offshore, as companies focus on core competencies and ways to cut costs. From an information security point of view, these arrangements can present risk that may be difficult to quantify and potentially difficult to mitigate. Typically, both the resources and skills of the outsourced functions are lost to the organization, which itself will present risk. Providers may operate on different standards and can be difficult to control. The security strategy should consider outsourced security services carefully to ensure that they either are not a critical single point of failure or that there is a viable backup plan in the event of service provider failure.[17]

Risk posed by outsourcing can also materialize as the result of mergers and acquisitions. Typically, significant differences in culture, systems, technology and operations between the parties present a host of security challenges that must be identified and addressed. Often, in these situations, security is an afterthought and the security manager must strive to gain a presence in these activities and assess the risk for management consideration.[18]

---

[17] ISACA, *CISM Review Manual 15th Edition*, USA, 2016
[18] *Ibid.*

TOPIC 2—COMMON ATTACK TYPES AND VECTORS

Attack vectors and methodologies continue to evolve, which represents a significant threat on the client side. Although some attacks are made at random with no particular target in mind, targeted attacks are made against recipients who have been researched and identified as useful by attackers. Phishing attacks are often directed toward recipients who have access to data or systems to which the attacker wishes to gain access. In other cases, malware is deployed in widespread attacks with the hope that it will hit as many vulnerable systems as possible, though these situations are not likened to cyberattacks. A number of distinct threat agents and attack patterns have emerged in the current threat landscape. It is essential for cybersecurity professionals to be able to identify these threats in order to manage them appropriately.

THREAT AGENTS

The European Union Agency for Network and Information Security (ENISA) has identified threat agents that currently exist in the threat landscape, shown in **figure 2.5**.

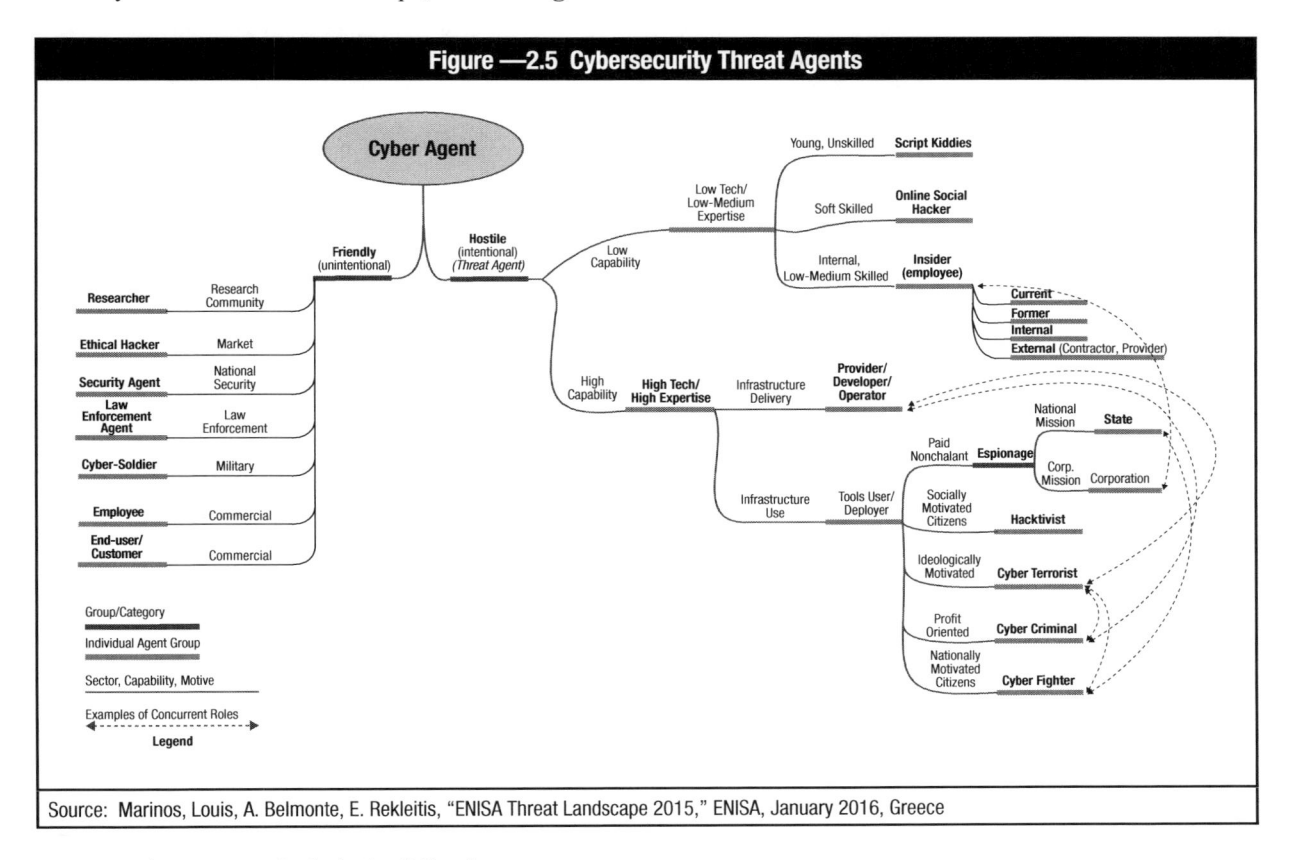

Figure —2.5 Cybersecurity Threat Agents

Source: Marinos, Louis, A. Belmonte, E. Rekleitis, "ENISA Threat Landscape 2015," ENISA, January 2016, Greece

Common threat agents include the following:
- **Corporations**—Corporations have been known to breach security boundaries and perform malicious acts to gain a competitive advantage.
- **Cybercriminals**—Motivated by the desire for profit, these individuals are involved in fraudulent financial transactions.
- **Cyberterrorists**—Characterized by their willingness to use violence to achieve their goals, cyberterrorists frequently target critical infrastructures and government groups.
- **Cyberwarriors**—Often likened to hacktivists, cyberwarriors, also referred to as cyberfighters, are nationally motivated citizens who may act on behalf of a political party or against another political party that threatens them.

- **Employees**—Although they typically have fairly low-tech methods and tools, dissatisfied current or former employees represent a clear cybersecurity risk. All of these attacks are adversarial, but some are not related to APT cyberattacks.
- **Hacktivists**—Although they often act independently, politically motivated hackers may target specific individuals or organizations to achieve various ideological ends.
- **Nation states**—Nation states often target government and private entities with a high level of sophistication to obtain intelligence or carry out other destructive activities.
- **Online social hackers**—Skilled in social engineering, these attackers are frequently involved in cyberbullying, identity theft and collection of other confidential information or credentials.
- **Script kiddies**—Script kiddies are individuals who are learning to hack; they may work alone or with others and are primarily involved in code injections and distributed denial-of-service (DDoS) attacks.

ATTACK ATTRIBUTES

While risk is measured by potential activity, an **attack** is the actual occurrence of a threat. More specifically, an attack is an activity by a threat agent (or adversary) against an asset. From an attacker's point of view, the asset is a **target**, and the path or route used to gain access to the target (asset) is known as an **attack vector**. There are two types of attack vectors: ingress and egress (also known as data exfiltration). While most attack analysis concentrates on ingress, or intrusion, hacking into systems, some attacks are designed to remove data (e.g., employees that steal data) from systems and networks. Therefore, it is important to consider both types of attack vectors.

The attacker must defeat any controls in place and/or use an **exploit** to take advantage of a vulnerability. Another attribute of an attack is the **attack mechanism**, or the method used to deliver the exploit. Unless the attacker is personally performing the attack, the attack mechanism may involve an exploit that delivers the payload to the target. An example can be a crafted malicious pdf, crafted by the attacker and delivered by email. The attributes of an attack are shown in **figure 2.6**.

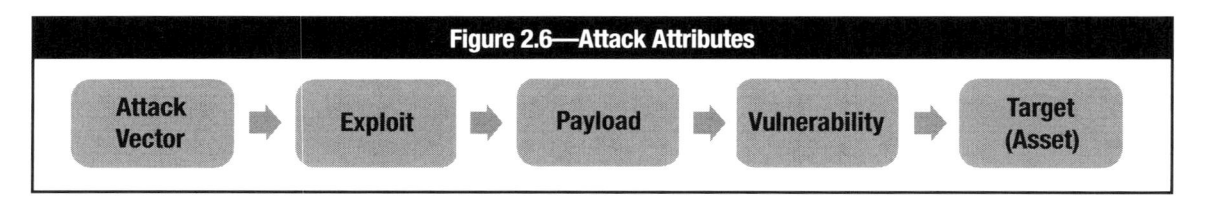

Figure 2.6—Attack Attributes

Attack Vector → Exploit → Payload → Vulnerability → Target (Asset)

Detailed analysis of cyberattacks requires significant technical and subject matter expertise and is an important part of cybersecurity. Each of the attack attributes (attack vector, exploit, payload, vulnerability, target) provides unique points where controls to prevent or detect the attack can be placed. It is also essential to understand each of these attributes when analyzing and investigating an actual attack. For example, the exploit used to deliver the payload often leaves artifacts or evidence that can be used by technical analysts and investigators to understand the attack and potentially identify the perpetrators. Analysis of the data exfiltration path may identify additional opportunities to prevent or detect the removal of data or obtain evidence, even if the attack was able to gain access to the target.

Attacks can be analyzed and categorized based on their type and patterns of use. From these characteristics, it is possible to make generalizations that facilitate better design and controls. There are two broad categories for threat events: adversarial and nonadversarial. An **adversarial threat event** is made by a human threat agent (or adversary), while a **nonadversarial threat event** is usually the result of an error, malfunction or mishap of some sort.[19]

---

[19] MITRE, Common Attack Pattern Enumeration and Classification (CAPEC), February 2014, *http://capec.mitre.org*

GENERALIZED ATTACK PROCESS

While each attack is different, most adversarial threat events follow a common process, as shown in **figure 2.7** and described below.

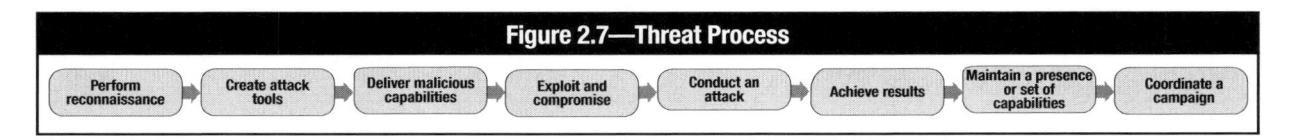

Figure 2.7—Threat Process

Perform reconnaissance → Create attack tools → Deliver malicious capabilities → Exploit and compromise → Conduct an attack → Achieve results → Maintain a presence or set of capabilities → Coordinate a campaign

1. **Perform reconnaissance:** The adversary gathers information using a variety of techniques, passive or active, which may include:
   a. Passive:
      i. Sniffing network traffic
      ii. Using open source discovery of organizational information (news groups; company postings on IT design and IT architecture)
      iii. Google hacking
   b. Active:
      i. Scanning the network perimeter
      ii. Social engineering (fake phone calls, low-level phishing)
2. **Create attack tools:** The adversary crafts the tools needed to carry out a future attack, which include:
   a. Phishing or spear phishing attacks
   b. Crafting counterfeit websites or certificates
   c. Creating and operating false organizations and placing them in to the supply chain to inject malicious components
3. **Deliver malicious capabilities:** The adversary inserts or installs whatever is needed to carry out the attack, which includes the following:
   a. Introducing malware into organizational information systems
   b. Placing subverted individuals into privileged positions within the organization
   c. Installing sniffers or scanning devices on targeted networks and systems
   d. Inserting tampered hardware or critical components into organizational systems or supply chains
4. **Exploit and compromise:** The adversary takes advantage of information and systems in order to compromise them, which include:
   a. Split tunneling or gaining physical access to organizational facilities
   b. Exfiltrating data or sensitive information
   c. Exploiting multitenancy (i.e., multiple customers on shared resources) in a public cloud environment (e.g., attacking open public access points; application program interfaces [APIs])
   d. Launching zero-day exploits
5. **Conduct an attack:** The adversary coordinates attack tools or performs activities that interfere with organizational functions. Potential methods of attack include:
   a. Communication interception or wireless jamming attacks
   b. Denial-of-service (DoS) or distributed DDoS attacks
   c. Remote interference with or physical attacks on organizational facilities or infrastructures
   d. Session-hijacking or man-in-the-middle attacks
6. **Achieve results:** The adversary causes an adverse impact, which may include:
   a. Obtaining unauthorized access to systems and/or sensitive information
   b. Degrading organizational services or capabilities
   c. Creating, corrupting or deleting critical data
   d. Modifying the control flow of information system (e.g., industrial control system, supervisory control and data acquisition (SCADA) systems)
7. **Maintain a presence or set of capabilities:** The adversary continues to exploit and compromise the system using the following techniques:
   a. Obfuscating adversary actions or interfering with intrusion detection systems (IDSs)
   b. Adapting cyberattacks in response to organizational security measures

Section 2: Cybersecurity Concepts

8. **Coordinate a campaign:** The adversary coordinates a campaign against the organization that may involve the following measures:
   a. Multi-staged attacks
   b. Internal and external attacks
   c. Widespread and adaptive attacks

NONADVERSARIAL THREAT EVENTS

Although most attacks are the result of a coordinated effort, there are other events that can pose various types of risk to an organization and can aid an adversary in a possible cyber-attack. Some of the most common nonadversarial threat events are:
- Mishandling of critical or sensitive information by authorized users
- Incorrect privilege settings
- Fire, flood, hurricane, windstorm or earthquake at primary or backup facilities
- Introduction of vulnerabilities into software products
- Pervasive disk errors or other problems caused by aging equipment

MALWARE, RANSOMWARE AND ATTACK TYPES[20]

Malware, also called malicious code, is software designed to gain access to targeted computer systems, steal information or disrupt computer operations. There are several types of malware, the most important being computer viruses, network worms and Trojan horses, which are differentiated by the way in which they operate or spread.

For example, the computer worm known as Stuxnet highlights malware's potential to disrupt SCADA systems and programmable logic controllers (PLCs), typically used to automate mechanical processes in factory settings or power plants. Discovered in 2010, Stuxnet was used to compromise Iranian nuclear systems and software. It has three components:
1. A **worm** that carries out routines related to the payload
2. A **link file** that propagates copies of the worm
3. A **rootkit** that hides malicious processes to prevent detection

Other common types of malware include:
- **Viruses**—A computer virus is a piece of code that can replicate itself and spread from one computer to another. It requires intervention or execution to replicate and/or cause damage.
- **Network worm**—A variant of the computer virus, which is essentially a piece of self-replicating code designed to spread itself across computer networks. It does not require intervention or execution to replicate.
- **Trojan horses**—A piece of malware that gains access to a targeted system by hiding within a genuine application. Trojan horses are often broken down into categories reflecting their purposes.
  - A common mobile Trojan is Hummer, a type of Android malware. In the first six months of 2016, nearly 1.4 million devices daily were infected by Hummer. The malware uses one of 18 rooting methods to gain root privileges to the device. It then pushes ads and installs games and pornographic applications on to the mobile device.[21]
- **Botnets**—Derived from "robot network," a large, automated and distributed network of previously compromised computers that can be simultaneously controlled to launch large-scale attacks such as DoS.

A number of further terms are also used to describe more specific types of malware, characterized by their purposes. They include:
- **Spyware**—A class of malware that gathers information about a person or organization without the knowledge of that person or organization.
- **Adware**—Designed to present advertisements (generally unwanted) to users.

---

[20] ISACA, *Advanced Persistent Threats: How to Manage the Risk to Your Business*, USA, 2013
[21] Bisson, David; "Hummer Malware the No. 1 Mobile Trojan in the World," *Tripwire*, 1 July 2016, www.tripwire.com/state-of-security/latest-security-news/hummer-malware-the-1-mobile-trojan-in-the-world

- **Ransomware**—Also called "hostage code," a class of extortive malware that locks or encrypts data or functions and demands a payment to unlock them. Several types are available for every operating system. Recent examples of ransomware attacks include:
  - **GhostCrypt**—Using AES encryption, GhostCrypt scrambles the data on the infected device in order to obtain Bitcoins from the victim. Whenever the ransomware encrypts any piece of information, it attaches the .Z81928819 appendix. GhostCrypt also generates a READ_THIS_FILE.txt as a reminder of the ransom to be paid.[22]
  - **SNSLocker**—Using AES encryption and adding a .RSNSlocked string at the end of affected data items, this ransomware encrypts data and demands a ransom of $300 (payable in Bitcoin) for the decryption solution.[23]
- **Keylogger**—A class of malware that secretly records user keystrokes and, in some cases, screen content.
- **Rootkit**—A class of malware that hides the existence of other malware by modifying the underlying operating system.

OTHER ATTACK TYPES

In addition to malware and ransomware, there are many other types of attacks. Some of the most common attack patterns are as follows:

- **Advanced persistent threats (APTs)**—Complex and coordinated attacks directed at a specific entity or organization. They require a substantial amount of research and time, often taking months or even years to fully execute. APT is a term, indicating the class of complexity; however, it cannot be tested if a particular attack was APT or not. After an attack is discovered and the level of complexity is determined and the amount of time and resources spent on the attack is investigated, the attack can be classified as an attack by an APT.
- **Backdoor**—A means of regaining access to a compromised system by installing software or configuring existing software to enable remote access under attacker-defined conditions.
- **Brute force attack**—An attack made by trying all possible combinations of passwords or encryption keys until the correct one is found.
- **Buffer overflow**—Occurs when a program or process tries to store more data in a buffer (temporary data storage area) than it was intended to hold. Since buffers are created to contain a finite amount of data, the extra information—which has to go somewhere—can overflow into adjacent buffers, corrupting or overwriting the valid data held in them. Although it may occur accidentally through programming error, buffer overflow is an increasingly common type of security attack on data integrity. In buffer overflow attacks, the extra data may contain codes type of security attack on data integrity.
- **Cross-site scripting (XSS)**—A type of injection in which malicious scripts are injected into otherwise benign and trusted websites. XSS attacks occur when an attacker uses a web application to send malicious code, generally in the form of a browser side script, to a different end user. Flaws that allow these attacks to succeed are quite widespread and occur anywhere a web application uses input from a user within the output it generates without validating or encoding it.
- **DoS attack**—An assault on a service from a single source that floods it with so many requests that it becomes overwhelmed and is either stopped completely or operates at a significantly reduced rate.
- **Man-in-the-middle attack**—An attack strategy in which the attacker intercepts the communication stream between two parts of the victim system and then replaces the traffic between the two components with the intruder's own, eventually assuming control of the communication.
- **Social engineering**—Any attempt to exploit social vulnerabilities to gain access to information and/or systems. It involves a "con game" that tricks others into divulging information or opening malicious software or programs.
- **Phishing**—A type of email attack that attempts to convince a user that the originator is genuine, but with the intention of obtaining information for use in social engineering.
- **Spear phishing**—An attack where social engineering techniques are used to masquerade as a trusted party to obtain important information such as passwords from the victim.
- **Spoofing**—Faking the sending address of a transmission in order to gain illegal entry into a secure system.

---

[22] Tripwire, "May 2016: The Month in Ransomware," *Tripwire*, 6 June 2016, *www.tripwire.com/state-of-security/security-data-protection/may-2016-the-month-in-ransomware*
[23] *Ibid.* [24] OWASP, SQL Injection, *www.owasp.org/index.php/SQL_Injection*

- **Structure Query Language (SQL) injection**—According to OWASP,[24] "A SQL injection attack consists of insertion or 'injection' of a SQL query via the input data from the client to the application. A successful SQL injection exploit can read sensitive data from the database, modify database data (Insert/Update/Delete), execute administration operations on the database (such as shutdown the DBMS), recover the content of a given file present on the DBMS file system and in some cases issue commands to the operating system. SQL injection attacks are a type of injection attack, in which SQL commands are injected into data-plane input in order to effect the execution of predefined SQL commands."
- **Zero-day exploit**—A vulnerability that is exploited before the software creator/vendor is even aware of its existence.

TOPIC 3—POLICIES

PURPOSE OF POLICIES

Information security policies are a primary element of cybersecurity and overall security governance. They specify requirements and define the roles and responsibilities of everyone in the organization, along with expected behaviors in various situations. Therefore, they must be properly created, accepted and validated by the board and senior management before being communicated throughout the organization. During this process, there may be occasions where other documents must be created to address unique situations separate from the bulk of the organization. This may be necessary when part of the organization has a specific regulatory requirement to protect certain types of information.

POLICY LIFE CYCLE

In addition to a policy framework, another important aspect of information security policies is their life cycle of development, maintenance, approval and exception.

Every compliance document should have a formal process of being created, reviewed, updated and approved at least once a year. Additionally, there may be legitimate need for an exception to a policy; therefore, a clear process of how an exception is approved by senior management and monitored during the life cycle is necessary.

GUIDELINES

There are several attributes of good policies that should be considered:
- Security policies should be an articulation of a well-defined information security strategy that captures the intent, expectations and direction of management.
- Policies must be clear and easily understood by all affected parties.
- Policies should be short and concise, written in plain language.

Most organizations should create security policies prior to developing a security strategy. Although many organizations tend to follow an *ad hoc* approach to developing security strategy, there are also instances, especially in smaller organizations, where effective practices have been developed that may not be reflected in written policies. Existing practices that adequately address security requirements may usefully serve as the basis for policy and standards development. This approach minimizes organizational disruptions, communications of new policies and resistance to new or unfamiliar constraints.

COMPLIANCE DOCUMENTS AND POLICY FRAMEWORKS

Compliance documents, such as policies, standards and procedures, outline the actions that are required or prohibited. Violations may be subject to disciplinary actions.

Some common compliance document types are shown in **figure 2.8**.

Figure 2.8—Compliance Document Types	
Type	**Description**
Policies	Communicate required and prohibited activities and behaviors
Standards	Interpret policies in specific situations
Procedures	Provide details on how to comply with policies and standards
Guidelines	Provide general guidance on issues such as "what to do in particular circumstances." These are not requirements to be met but are strongly recommended.

Some organizations may not implement all of these types of documents. For example, smaller organizations may simply have policies and procedures; others may have policies, standards and procedures, but not guidelines.

TYPES OF INFORMATION SECURITY POLICIES

The number and type of policies an organization chooses to implement varies based on the organization's size, culture, risk, regulatory requirements and complexity of operations. However, following are some common examples and the type of information they might contain.[25]

General Information Security Policy

Most organizations have a general, high-level information security policy that may stand alone as a single policy or serve as a foundation for other compliance documents. For larger enterprises, it is common practice to subdivide policies by topic to address all of the information security. An example of such a subdivision is shown in **figure 2.9**.

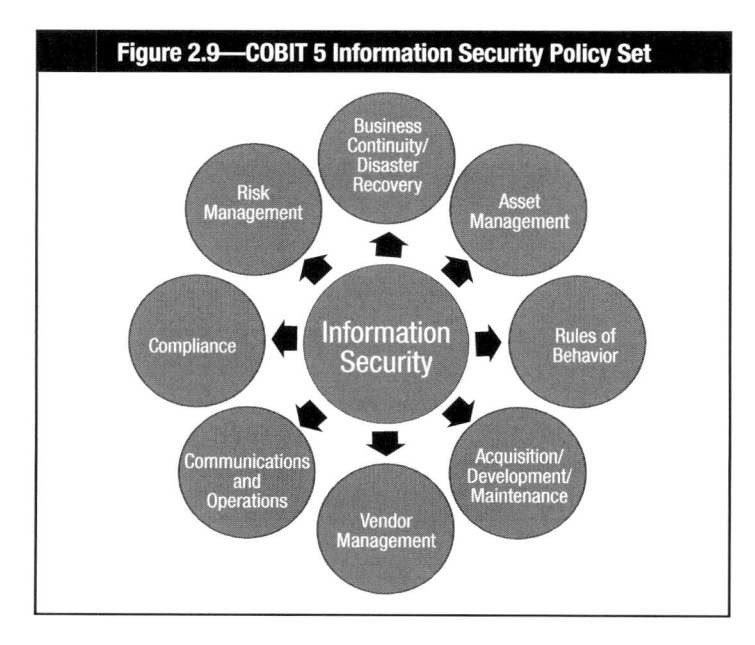

Figure 2.9—COBIT 5 Information Security Policy Set

Each of these policies requires the input of information security. Examples for a possible relevant scope for information security are as follows:[26]
- Business Continuity and Disaster Recovery:
  – Business impact analysis (BIA)
  – Business contingency plans with trusted recovery
  – Recovery requirements for critical systems
  – Defined thresholds and triggers for contingencies and escalation
  – Disaster recovery plan (DRP)
  – Training and Testing
- Asset Management:
  – Data classification and ownership
  – System classification and ownership
  – Resource utilization and prioritization
  – Asset life cycle management
  – Asset protection
- Rules of Behavior:
  – At-work acceptable use and behavior, including privacy, Internet/email, mobile devices, BYOD, etc.
  – Offsite acceptable use and behavior, including social media, blogs

---

[25] ISACA, *COBIT® 5 for Information Security*, USA, 2013
[26] *Ibid.*

- Acquisition/Development/Maintenance:
  - Information security within the life cycle, requirements definition and procurement/acquisition processes
  - Secure coding practices
  - Integration of information security with change and configuration management
- Vendor Management:
  - Contract management
- Communication and Operations:
  - IT information security architecture and application design
  - Service level agreements
- Compliance:
  - IT information security compliance assessment process
  - Development of metrics
  - Assessment repositories
- Risk Management:
  - Organizational risk management plan
  - Information risk profile

The appearance and length of an information security policy varies greatly among enterprises. Some enterprises consider a one-page overview to be a sufficient information security policy. In these cases, the policy could be considered a directive statement, and it should clearly describe links to other specific policies. In other enterprises, the information security policy is fully developed, containing nearly all the detailed guidance needed to put the principles into practice. It is important to understand what the information stakeholders expect in terms of coverage and to adapt to this expectation.

Regardless of its size or degree of detail, the information security policy needs a clearly defined scope. This involves:
- The enterprise's definition of information security
- The responsibilities associated with information security
- The vision for information security, accompanied by appropriate goals, metrics and rationale of how the vision is supported by the information security culture and awareness
- Explanation of how the information security policy aligns with other high-level policies
- Elaboration on specific information security topics such as data management, information risk assessment and compliance with legal, regulatory and contractual obligations

In addition to the elements discussed above, a policy may potentially affect the security life cycle budget and cost management. Information security strategic plans and portfolio management can be added as well.

The policy should be actively communicated to the entire enterprise and distributed to all employees, contractors, temporary employees and third-party vendors. Stakeholders need to know the information principles, high-level requirements, and roles and responsibilities for information security. The responsibility for updating and revalidating the information security policy lies with the cybersecurity function.

Other possible security policies or procedures include access control, personnel information and security incidents.

Access Control Policy
The access control policy provides proper access to internal and external stakeholders to accomplish business goals. This can be measured by metrics such as, but not limited to, the:
- Number of access violations that exceed the amount allowed
- Amount of work disruption due to insufficient access rights
- Number of segregation of duties incidents or audit findings

Additionally, the access control policy should ensure that emergency access is appropriately permitted and revoked in a timely manner. Metrics related to this goal include the number of emergency access requests and the number of active emergency accounts in excess of approved time limits.

The access control policy should cover the following topics, among others:
- Physical and logical access provisioning life cycle
- Least privilege/need to know
- Segregation of duties
- Emergency access

This policy is meant for all corresponding business units, vendors and third parties. Updates and revalidation should involve HR, data and system owners, information security and senior management. A new or updated policy should be distributed to all corresponding business units, vendors and third parties.

Personnel Information Security Policy

The personnel information security policy objective includes, but is not limited to, the following goals:
- Execute regular background checks of all employees and people at key positions. This goal can be measured by counting the number of completed background checks for key personnel. This can be amplified with the number of overdue background check renewals based on a predetermined frequency.
- Acquire information about key personnel in information security positions. This can be followed up by counting the number of personnel in key positions that have not rotated according to a predefined frequency.
- Develop a succession plan for all key information security positions. A starting point is to list all the critical information security positions that lack backup personnel.
- Define and implement appropriate procedures for termination. This should include details about revoking account privileges and access.

This policy is meant for all corresponding business units, vendors and third parties. Updates and revalidation should involve HR, the privacy officer, the legal department, information security and facility security. A new or updated policy needs to be distributed to employees, contract personnel, vendors under contract and temporary employees.

Security Incident Response Policy

This policy addresses the need to respond to (cybersecurity) incidents in a timely manner in order to recover business activities. The policy should include:
- A definition of an information security incident
- A statement of how incidents will be handled
- Requirements for the establishment of the incident response team, with organizational roles and responsibilities
- Requirements for the creation of a tested incident response plan, which will provide documented procedures and guidelines for:
  - Criticality of incidents
  - Reporting and escalation processes
  - Recovery (including):
    - Recovery point objectives (RPOs): The RPO is determined based on the acceptable data loss in case of disruption of operations. It indicates the most recent point in time to which it is acceptable to recover the data, which generally is the latest backup. RPO effectively quantifies the permissible amount of data loss in case of interruption. Depending on the volume of data, it may be advisable to reduce the time between backups to prevent a situation where recovery becomes impossible because of the volume of data to be restored. It may also be the case that the time required to restore a large volume of data makes it impossible to achieve the RTO.[27]
    - Recovery time objectives (RTOs) for return to the trusted state, including:
      - Investigation and preservation of process
      - Testing and training
      - Post incident meetings to document root cause analysis and enhancements of information security practices that prevent similar future events
        - Incident documentation and closing

---

[27] ISACA, *CISM Review Manual 15th Edition*, USA, 2016

This policy is meant for all corresponding business units and key employees. Updates and revalidation should involve the information security function. A new or updated policy should be distributed to key employees.

Policy Frameworks

The way that compliance documents relate to and support each other is called a policy framework. A framework defines different types of documents and what is contained in each. Organizations may have simple or relatively complex policy frameworks depending on their unique needs. Organizations may define a separate cybersecurity policy, but this should always be part of the overarching information security policy framework.

Page intentionally left blank

TOPIC 4—CYBERSECURITY CONTROLS

Cybersecurity is a dynamic and ever-changing environment and requires continuous monitoring, updating, testing, patching and changing as technology and business evolve. These controls are critical to maintaining security within any organization's IT infrastructure. Failure to address these processes is one of the top causes of security breaches in organizations.

An excellent resource for gaining more in-depth knowledge on cybersecurity controls is the Center for Internet Security (CIS) Critical Security Controls for Effective Defense. It provides actionable guidance to stop the most pervasive and dangerous attacks in the current environment. The CIS Critical Security Controls are derived from common attack patterns as provided by leading threat reports from a wide community of industry practitioners. They provide an organized means for cybersecurity professionals to address these common threats and attacks.[28]

IDENTITY MANAGEMENT

Cybersecurity relies upon the establishment and maintenance of user profiles that define the authentication, authorization and access controls for each user. Today, organizations have a variety of *ad hoc* processes and tools to manage and provision user identity information. Identity management focuses on streamlining various business processes needed to manage all forms of identities in an organization—from enrollment to retirement.

The ability to integrate business processes and technology is critically important in the emerging model because it links people to systems and services. A key objective of identity management is to centralize and standardize this process so that it becomes a consistent and common service across the organization.

Identity management is comprised of many components that provide a collective and common infrastructure, including directory services, authentication services (validating who the user is) and authorization services (ensuring the user has appropriate privileges to access systems based on a personalized profile). It also includes user-management capabilities, such as user provisioning and deprovisioning, and can include the utilization of federated identity management (FIM).

FIM allows a user from one business entity to seamlessly access resources of another business entity in a secure and trustworthy manner. Federated single sign-on (SSO) between the issuing domain (identity provider) and a relying domain (service provider) facilitates the secure and trusted transfer of user identifiers and other attributes. FIM also supports standards-based trust and security for applications exposed as web services.

PROVISIONING AND DEPROVISIONING

User provisioning is part of the organization's hiring process where user accounts are created. Passwords and access control rights are generally assigned based on the job duties of the users. This can be a complicated process, as users may need access to many different resources such as systems, databases, email, applications and remote services, each of which has its own access control, passwords, encryption keys or other authorization and authentication requirements. Additionally, access control rights often change based on shifting job requirements, so it is frequently necessary to update access controls and remove access that is no longer needed. Likewise, when a user leaves an organization, their accounts need to be deprovisioned—meaning that all accounts and accesses must be suspended or deleted in a timely manner.

AUTHORIZATION[29]

The authorization process used for access control requires that the system be able to identify and differentiate among users. Access rules (authorizations) specify who can access what. For example, access control is often based on least privilege, which means granting users only those accesses required to perform their duties. Access should be on a documented need-to-know and need-to-do basis by type.

---

[28] Center for Internet Security (CIS), *The CIS Critical Security Controls for Effective Cyber Defense*, www.sans.org/critical-security-controls
[29] ISACA, *CISA Review Manual 26th Edition*, USA, 2015

Computer access can be set for various levels (e.g., files, tables, data items, etc.). When IS auditors review computer accessibility, they need to know what can be done with the access and what is restricted. For example, access restrictions at the file level generally include the following:
- Read only
- Write, create, update only
- Delete only
- Execute only
- A combination of the above

The least dangerous type of access is read-only, as long as the information being accessed is not sensitive or confidential. This is because the user cannot alter or use the computerized file beyond basic viewing or printing.

ACCESS CONTROL LISTS[30]

To provide security authorizations for the files and facilities listed previously, logical access control mechanisms utilize access authorization tables, also referred to as access control lists (ACLs) or access control tables. ACLs refer to a register of:
- Users (including groups, machines, processes) who have permission to use a particular system resource
- The types of access permitted

ACLs vary considerably in their capability and flexibility. Some only allow specifications for certain preset groups (e.g., owner, group and global), while more advanced ACLs allow much more flexibility such as user-defined groups. Also, more advanced ACLs can be used to explicitly deny access to a particular individual or group. With more advanced ACLs, access can be at the discretion of the policy maker (and implemented by the security administrator) or individual user, depending upon how the controls are technically implemented. When a user changes job roles within an organization, often their old access rights are not removed before adding their new required accesses. Without removing the old access rights, there could be a potential segregation of duties issue.

ACCESS LISTS[31, 32]

Access lists filter traffic at network interfaces based on specified criteria, thus affording basic network security. Without access lists, network devices pass all packets. Conversely, after an access list is created and applied to an interface, it then only passes traffic permitted by rules due to an implied "deny all" statement automatically appended to the list. Understanding the placement and impact of an access list is essential because errors can halt network traffic entirely.

PRIVILEGED USER MANAGEMENT

Privileged access permits administrators to maintain and protect systems and networks. Privileged users can often access any information stored within a system, which means they can modify or circumvent existing safeguards such as access controls and logging. "Privileged user" typically refers to the administrators of systems, networks, servers or workstations.

Because of this elevated access, organizations need to think carefully about privileged users and accounts and apply additional controls to them. Common controls include:
- Limiting privileged access to only those who require it to perform their job functions
- Performing background checks on individuals with elevated access
- Implementing additional logging of activity associated with privileged accounts
- Maintaining accountability by never sharing privileged accounts
- Using stronger passwords or other authentication controls to protect privileged accounts from unauthorized access
- Regularly reviewing accounts for privileges and removing those no longer required
- Require privileged users to maintain two accounts (elevated and non-elevated) and mandate the use of non-elevated access accounts to general duties, such as email, documenting, accessing the Internet, etc.

---

[30] Ibid.
[31] Wilson, Tracey, "Basics of Access Control Lists: How to Secure ACLs," 16 May 2012, http://blog.pluralsight.com/access-control-list-concepts
[32] Cisco; Access Control Lists: Overview and Guidelines, Cisco IOS Security Configuration Guide, www.cisco.com/c/en/us/td/docs/ios/12_2/security/configuration/guide/fsecur_c/scfacls.pdf

CHANGE MANAGEMENT

Change management is essential to the IT infrastructure. Its purpose is to ensure that that changes to processes, systems, software, applications, platforms and configuration are introduced in an orderly, controlled manner. Controls are implemented in the form of a structured review process intended to evaluate and minimize the potential for disruption that a proposed change, maintenance activity or patch may introduce. Effective controls ensure that all changes are categorized, prioritized and authorized. The process generally includes mechanisms for tracking and documenting changes to demonstrate accountability and compliance with best practices.

It is important to note that change management is not a stand-alone process; it draws upon a number of other processes and controls. Therefore, it requires a comprehensive knowledge of enterprise operations and infrastructure to be implemented effectively.

CONFIGURATION MANAGEMENT

Maintaining the security configurations of network devices, systems, applications and other IT resources is critically important to ensure security controls are properly installed and maintained. As organizations grow and evolve, so does the potential for change and dysfunction. In order to manage such changes and minimize their potential to disrupt operations, efficiency and profits, it is necessary to develop formal processes. These processes of configuration management can be quite complex, as they support many other activities within the enterprise.

Implementing a configuration management process has several benefits for security including:[33]
- Verification of the impact on related items
- Assessment of a proposed change's risk
- Ability to inspect different lines of defense for potential weaknesses
- Tracking of configuration items against approved secure configuration baselines
- Insights into investigations after a security breach or operations disruption
- Version control and production authorization of hardware and software components

PATCH MANAGEMENT

Patches are solutions to software programming errors. In many cases, security vulnerabilities are introduced by coding errors. Therefore, it is vital that software bugs that are identified as security vulnerabilities be patched as soon as possible. Most software vendors release regular software updates and patches as the vulnerabilities are identified and fixed.

Failure to apply patches to known security vulnerabilities is the most common cause of security breaches. Therefore, patching is an important part of vulnerability management, and organizations must set up processes to identify patches that are relevant to their IT infrastructure. Once a necessary patch is identified, it should be tested to ensure it does not negatively impact operations. After the patch has been verified, it can be scheduled and installed where appropriate.

---

[33] ISACA, *Configuration Management Using COBIT 5*, USA, 2013

Section 2: Cybersecurity Concepts

SECTION 2—KNOWLEDGE CHECK

Directions: Select the correct answer to complete each statement below. Use each word only once.

WORD BANK

Asset	Patches	Rootkit
Attack vector	Payload	Standards
Guidelines	Policies	Threat
Identity management	Procedure	Vulnerability
Malware	Cyberrisk	

1. The core duty of cybersecurity is to identify, mitigate and manage _____ to an organization's digital assets.
2. A(n) _____ is anything capable of acting against an asset in a manner that can cause harm.
3. A(n) _____ is something of value worth protecting.
4. A(n) _____ is a weakness in the design, implementation, operation or internal controls in a process that could be exploited to violate the system security.
5. The path or route used to gain access to the target asset is known as a(n) _____ .
6. In an attack, the container that delivers the exploit to the target is called a(n) _____ .
7. _____ communicate required and prohibited activities and behaviors.
8. _____ is a class of malware that hides the existence of other malware by modifying the underlying operating system.
9. _____ provide details on how to comply with policies and standards.
10. _____ provide general guidance and recommendations on what to do in particular circumstances.
11. _____, also called malicious code, is software designed to gain access to targeted computer systems, steal information or disrupt computer operations.
12. _____ are used to interpret policies in specific situations.
13. _____ are solutions to software programming and coding errors.
14. _____ includes many components such as directory services, authentication and authorization services, and user management capabilities such as provisioning and deprovisioning.

See answers in Appendix C.

Section 3: Security Architecture Principles

Topics covered in this section include:
1. Overview of security architecture
2. The OSI model
3. Defense in depth
4. Information flow control
5. Isolation and segmentation
6. Logging, monitoring and detection
7. Encryption fundamentals, techniques and applications

Page intentionally left blank

TOPIC 1—OVERVIEW OF SECURITY ARCHITECTURE

Security architecture describes the structure, components, connections and layout of security controls within an organization's IT infrastructure. Organizations have different types of security architectures that determine the particulars of various subsystems, products and applications. These particulars will in turn influence an organization's approach to **defense in depth**, or the practice of layering defenses to provide added protection.

Security architecture shows how defense in depth is implemented, as well as how layers of control are linked. Therefore, it is essential to designing and implementing security controls in any complex environment.

Each component of a given system poses its own security risk. Because the topology of security architecture varies from one organization to another, a number of different variables and risk should be considered when addressing the topology of a particular organization. This section will discuss those variables individually, along with best practices for successfully managing their related risk.

THE SECURITY PERIMETER

Many current security controls and architectures were developed with the concept of a perimeter—a well-defined (if mostly virtual) boundary between the organization and the outside world. In these models of cybersecurity, the focus is **network-** or **system-centric**. In the system-centric model, the emphasis is on placing controls at the network and system levels to protect the information stored within. An alternative model is **data-centric**, which emphasizes the protection of data regardless of its location.

With the advent of the Internet, outsourcing, mobile devices, cloud and other hosted services, the perimeter has expanded considerably. Consequently, there are significant new risk and vulnerabilities to confront in this hyper-connected and extended environment. The perimeter, then, is an important line of defense that protects the enterprise against external threats, and its design should reflect a proactive stance toward preventing potential risk.

An important component of the security perimeter is the Internet perimeter. This perimeter ensures secure access to the Internet for enterprise employees and guest users residing at all locations, including those involved in telecommuting or remote work. In order to provide security of email, front-end mobile and web apps, domain name system (DNS), etc., the Internet perimeter should:
- Route traffic between the enterprise and the Internet
- Prevent executable files from being transferred through email attachments or web browsing
- Monitor internal and external network ports for rogue activity
- Detect and block traffic from infected internal end point
- Control user traffic bound toward the Internet
- Identify and block anomalous traffic and malicious packets recognized as potential attacks
- Eliminate threats such as email spam, viruses and worms
- Enforce filtering policies to block access to websites containing malware or questionable content

The perimeter should also provide protection for virtual private networks (VPNs), wide area networks (WANs) and wireless local area networks (WLANs).

For VPNs, this protection should be threefold:
1. Terminate encrypted VPN traffic initiated by remote users.
2. Provide a hub for terminating encrypted VPN traffic from remote sites, organizations.
3. Provide a hub for terminating traditional dial-in users.

INTERDEPENDENCIES

As previously discussed, modern IT architectures are usually decentralized and deperimeterized. This includes a growing number of cloud-based platforms and services, as well as a shift in computing power and utilization patterns toward intelligent mobile devices such as tablets or smartphones. As a consequence, both the number of potential attack targets outside the organizational boundary and the number of attack vectors have grown. Conversely, the degree of control over deperimeterized environments has been significantly reduced, especially in enterprises permitting partial or full integration of user-owned mobile devices (i.e., BYOD). These changes have important ramifications for security architecture.

In distributed and decentralized IT architectures, the third-party risk is likely to increase, often as a function of moving critical applications, platforms and infrastructure elements into the cloud. For platforms, storage infrastructure and cloud-based data repositories, the focus of cybersecurity is shifting toward contracts and service level agreements (SLAs). Simultaneously, third-party cloud providers are facing an increased risk of attacks and breaches due to the agglomeration and clustering of sensitive data and information. In addition to concerns about third-party services, there is significant legal risk. Enterprises experiencing a loss of sensitive data may not be in a position to bring an action against the perpetrators because the cloud provider often has to initiate legal action.

Regardless of the generic information security arrangements made by an enterprise, there are often exposed areas within IT architectures. Cybercrime and cyberwarfare perpetrators continue to aim at "weak spots" in architectural elements and systems. In contrast to indiscriminate and opportunistic attacks, APTs and cybercrime always rely on preparatory research and insight into the target enterprise. This, in turn, raises the level of exposure for weak or unsecured parts of the overall architecture. These vulnerable spots include legacy systems, unpatched parts of the architecture, shared use of mobile devices and many others.

SECURITY ARCHITECTURES AND FRAMEWORKS

A great number of architectural approaches currently exist, and many of them have evolved from the development of enterprise architecture. Although their specific details may differ, they all generally aim to articulate what processes a business performs and how those processes are executed and secured. They articulate the organization, roles, entities and relationships that exist or should exist to perform a set of business processes.

Similarly, models of security architecture typically fall into two categories: process models and framework models. Frameworks allow a great deal of flexibility in how each element of the architecture is developed. The essence of a framework is to describe the elements of architecture and how they relate to one another, while a process model is more directive in its approach to the processes used for the various elements. A recent example of a process model is a web server building block where it is exactly specified how a web server should be deployed and what processing is and is not allowed within that block.

SABSA AND THE ZACHMAN FRAMEWORK

Just as there are many different types of business enterprises, there are many different approaches to security architecture. For example, the Zachman framework approach of developing a who, what, why, where, when and how matrix is shared by Sherwood Applied Business Security Architecture (SABSA). The matrix contains columns showing aspects of the enterprise that can be described or modeled, while the rows represent various viewpoints from which those aspects can be considered. This approach provides a logical structure for classifying and organizing design elements, which improves the completeness of security architecture.

THE OPEN GROUP ARCHITECTURE FRAMEWORK (TOGAF)

Another architecture framework is The Open Group Architecture Framework (TOGAF). Developed by The Open Group in the 1990s, this high-level and holistic approach addresses security as an essential component of the overall enterprise design. TOGAF's objective is to ensure that architectural development projects meet business objectives, that they are systematic and that their results are repeatable. **Figure 3.1** depicts the TOGAF architectural process and its relationship to businesses operations.

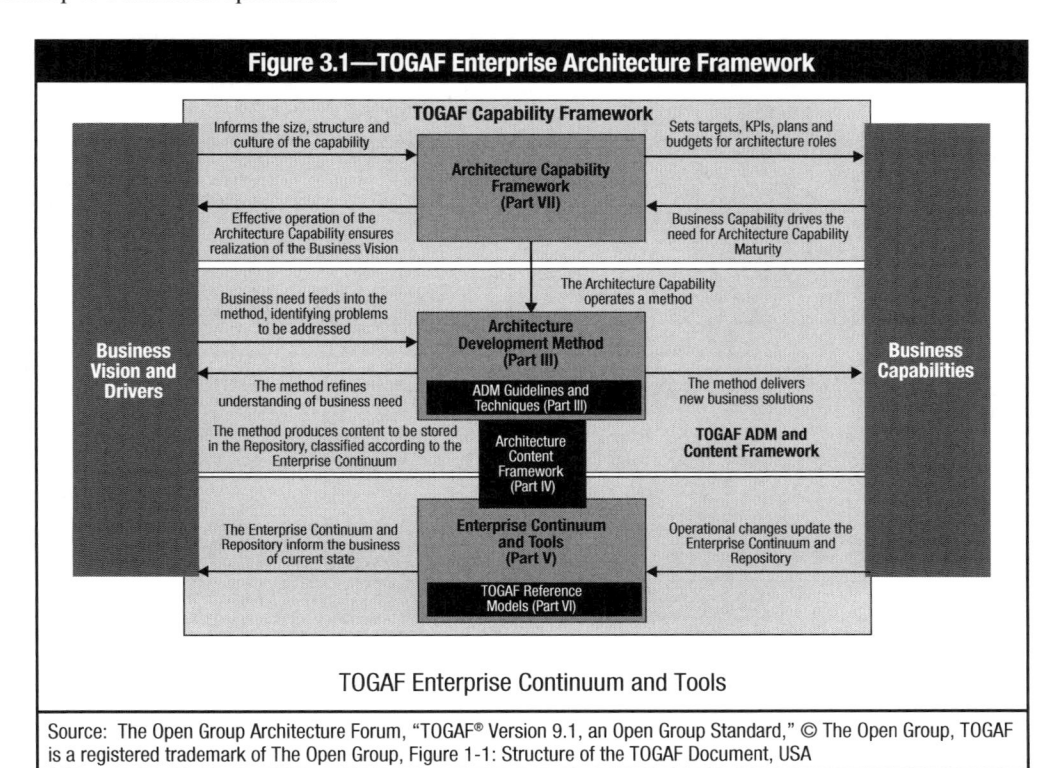

Figure 3.1—TOGAF Enterprise Architecture Framework

Source: The Open Group Architecture Forum, "TOGAF® Version 9.1, an Open Group Standard," © The Open Group, TOGAF is a registered trademark of The Open Group, Figure 1-1: Structure of the TOGAF Document, USA

Page intentionally left blank

TOPIC 2—THE OSI MODEL

The Open Systems Interconnection (OSI) model is used to describe networking protocols. Because it is rarely implemented in actual networks, it is considered a reference to standardize the development of actual networks. OSI was the first nonproprietary open definition for networking.

The OSI model defines groups of functionality required for network computers into layers, with each layer implementing a standard protocol for its functionality. There are seven layers in the OSI model, shown in **figure 3.2**.

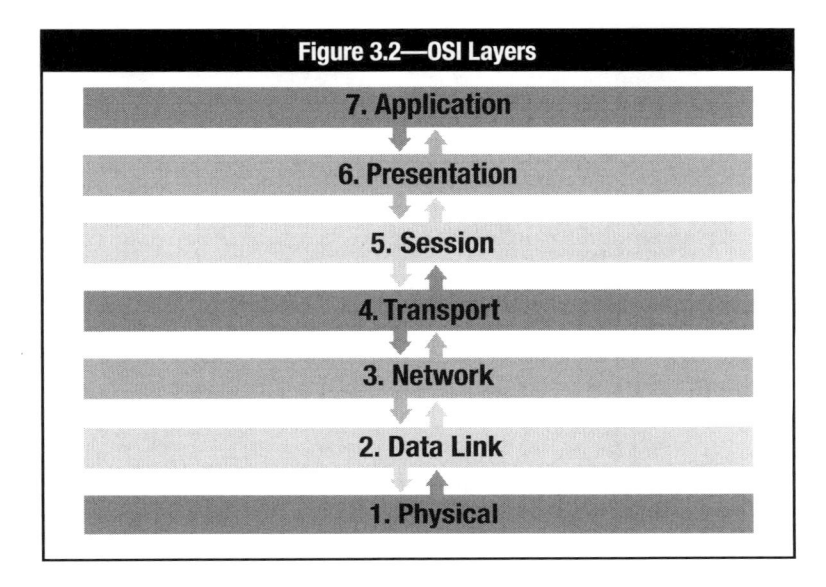

Figure 3.2—OSI Layers

7. Application
6. Presentation
5. Session
4. Transport
3. Network
2. Data Link
1. Physical

Each OSI layer performs a specific function for the network:
- **Physical layer (Layer 1)**—Manages signals among network systems
- **Data link layer (Layer 2)**—Divides data into frames that can be transmitted by the physical layer
- **Network layer (Layer 3)**—Translates network addresses and routes data from sender to receiver
- **Transport layer (Layer 4)**—Ensures that data are transferred reliably in the correct sequence
- **Session layer (Layer 5)**—Coordinates and manages user connections
- **Presentation layer (Layer 6)**—Formats, encrypts and compresses data
- **Application layer (Layer 7)**—Mediates between software applications and other layers of network services

TCP/IP

The protocol suite used as the de facto standard for the Internet is known as the Transmission Control Protocol/Internet Protocol (TCP/IP). The TCP/IP suite includes both network-oriented protocols and application support protocols and operates at Layer 3 and Layer 4 of the OSI model.

Internet Protocol Versions

Currently, there are two versions if IP that operate at Layer 3—IPv4 and IPv6.

IPv4 is the fourth revision of IP and is the most common IP used to connect devices to the Internet. It uses a 32-bit address scheme that allows for just over 4 billion addresses. With the current prevalence of Internet-connected devices, it is expected that IPv4 will eventually run out of unused addresses. Because of this, IPv6 has been developed to address this concern.[34]

IPv6, also called IPng (next generation) is the newest version of IP and is an evolutionary upgrade of IPv4. IPv6 was created to allow for the steady growth of the Internet for both the number of hosts connected and the amount of data transmitted. It is expected that IPv4 and IPv6 will coexist for some time.[35]

---
[34] Beal, Vangie, "What is the Difference Between IPv6 and IPv4?", Webopedia, 22 January 2014, *www.webopedia.com/DidYouKnow/Internet/ipv6_ipv4_difference.html*
[35] *Ibid.*

Section 3: Security Architecture Principles

Figure 3.3 shows some of the standards associated with the TCP/IP suite and where these fit within the OSI model. It is interesting to note that the TCP/IP set of protocols was developed before the OSI framework; therefore, there is no direct match between the TCP/IP standards and the layers of the framework.

Figure 3.3—OSI Association With the TCP/IP Suite

	OSI Model	TCP/IP Conceptual Layers	Protocol Data Unit (PDU)	TCP/IP Protocols	Equipment	Layer Functions	Layer Functions
7	Application	Application	Data	HTTP File Transport Protocol (FTP) Simple Mail Transport Protocol (SMTP) TFTP NFS Name Server Protocol (NSP) Simple Network Management Protocol (SNMP) Remote Terminal Control Protocol (Telnet) LPD X Windows DNS DHCP/BootP	Gateway	Provides user interface	File, print, message, database, and application services
6	Presentation					Presents data Handles processing such as encryption	Data encryption, compression and translation services
5	Session					Keeps separate the data of different applications	Dialog control
4	Transport	Transport	Segment	Transmission Control Protocol (TCP) User Datagram Protocol (UDP)	Layer 4 switch	Provide reliable or unreliable delivery	End-to-end connection
3	Network	Network interface	Packet	ICMP ARP RARP Internet Protocol (IP)	Router Layer 3 switch	Provides logical addressing which routers use for path determination	Routing
2	Data link	LAN or WAN interface	Frame	Ethernet Fast Ethernet FDDI Token Ring Point-to-point Protocol (PPP)	Layer 2 switch Bridge Wireless AP NIC	Combines packets into bytes and bytes into frames Provides access to media using MAC address Performs error detection, not error correction	Framing
1	Physical		Bits		Hub Repeater NIC	Moves bits between devices Specifies voltage, wire speed and pin-out of cables	Physical topology

Source: ISACA, *CISA Review Manual 26th Edition*, USA, 2015, figure 4.23

ENCAPSULATION

Encapsulation is the process of adding addressing information to data as they are transmitted down the OSI stack. Each layer relies on the services provided by the layer below. Each layer of the OSI model only communicates with its destination peer. It does so using datagrams or Protocol Data Units (PDUs). Refer to the previous **figure 3.3** for PDU names.

The OSI model is shown in **figure 3.4**. Upper layer data are passed down to the transport layer as segments and are "wrapped" with a header for identification. These segments are passed down to the network layer as packets again with a header. Data are broken down to frames at the data link layer and also have control information appended. At the physical layer, data take the form of bits (1s and 0s) for delivery to destination network.

Once at the destination, each layer on the receiving end strips off the appropriate addressing information and passes it up the OSI stack until the message is delivered. This process is called decapsulation.

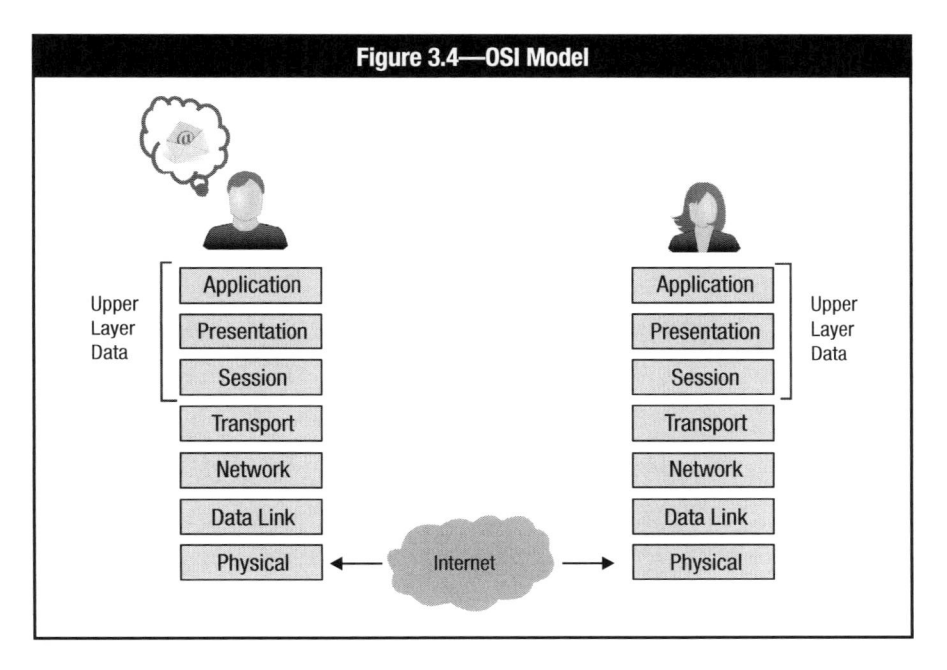

Figure 3.4—OSI Model

Communication services at OSI Layer 4 are categorized as either connection-oriented or connectionless. TCP provides reliable, sequenced delivery with error-checking. Connections are established using a three-way handshake, and thus are connection-oriented, as shown in **figure 3.5**. User Datagram Protocol (UDP) is a connectionless protocol used where speed is more important than error-checking and guaranteed delivery. UDP does use checksums for data integrity.

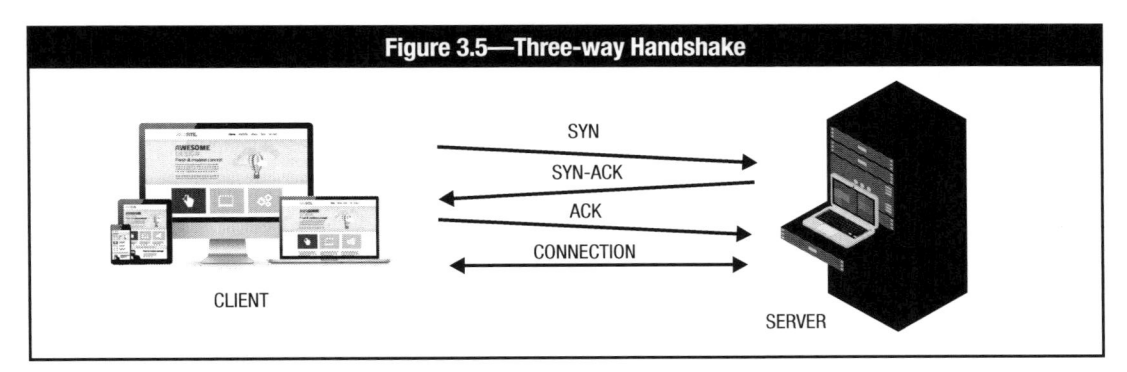

Figure 3.5—Three-way Handshake

Page intentionally left blank

TOPIC 3—DEFENSE IN DEPTH

Because no single control or countermeasure can eliminate risk, it is often important to use several controls to protect an asset. This process of layering defenses is known as **defense in depth**, but it may also be called protection in depth or security in depth. It forces an adversary to defeat or avoid more than one control to gain access to an asset.

Defense in depth is an important concept in designing an effective information security strategy or architecture. When designed and implemented correctly, multiple control layers provide multiple opportunities for monitoring to detect the attack. Adding additional controls to overcome also creates a delay so that the attack may be interrupted and prevented.

The number and types of layers needed are a function of asset value, criticality, the reliability of each control and the degree of exposure. Excessive reliance on a single control is likely to create a false sense of confidence. For example, a company that depends solely on a firewall can still be subject to numerous attack methodologies. A further defense may be to use education and awareness to create a "human firewall," which can constitute a critical layer of defense. Segmenting the network can constitute yet another defensive layer.

Using a defense in depth strategy for implementing controls has several advantages, including increasing the effort required for a successful attack and creating additional opportunities to detect or delay an attacker. There are several ways defense in depth can be implemented, as shown in **figure 3.6**.[36]

Figure 3.6—Types of Defense in Depth Implementations		
Type of Defense	**Graphical Representation**	**Description**
Concentric Rings (or nested layering)	Third layer of defense (Respond); Second layer of defense (Delay); First layer of defense (Detect)	Creates a series of nested layers that must be bypassed in order to complete an attack. Each layer delays the attacker and provides opportunities to detect the attack.
Overlapping redundancy	Control 1, Control 2, Control 3	Two or more controls that work in parallel to protect an asset. Provides multiple, overlapping points of detection. This is most effective when each control is different.

[36] Encurve, LLC.

Section 3: Security Architecture Principles

Figure 3.6—Types of Defense in Depth Implementations *(cont.)*

Type of Defense	Graphical Representation	Description
Segregation or compartmentalization		Compartmentalizes access to an asset, requiring two or more processes, controls or individuals to access or use the asset. This is effective in protecting very high value assets or in environments where trust is an issue.

Source: Encurve, LLC.

Another way to think about defense in depth is from an architectural perspective:
- **Horizontal defense in depth**—Controls are placed in various places in the path of access for an asset, which is functionally equivalent to concentric ring model shown in **figure 3.6**
- **Vertical defense in depth**—Controls are placed at different system layers—hardware, operating system, application, database or user levels

Using defense-in-depth techniques requires effective planning and understanding of each type's strengths and weaknesses as well as how the controls interact. It is easy to create an overly complex system of controls, and too many layers can be as bad as too few. When developing defense-in-depth implementations, consider the following questions:
- What vulnerabilities are addressed by each layer or control?
- How does the layer mitigate the vulnerability?
- How does each control interact with or depend on the other controls?

TOPIC 4—INFORMATION FLOW CONTROL

The Internet's openness makes every corporate network connected to it vulnerable to attack. Hackers on the Internet could break into a corporate network and do harm in a number of ways—by stealing or damaging important data, by damaging individual computers or the entire network, by using the corporate computer's resources or by using the corporate network and resources to pose as a corporate employee. Firewalls are built as one means of perimeter security for these networks.

A **firewall** is defined as a system or combination of systems that enforces a boundary between two or more networks, typically forming a barrier between a secure and an open environment such as the Internet. It applies rules to control the type of networking traffic flowing in and out. Most commercial firewalls are built to handle commonly used Internet protocols.

Effective firewalls should allow individuals on the corporate network to access the Internet and simultaneously prevent others on the Internet from gaining access to the corporate network to cause damage. Most organizations follow a deny-all philosophy, which means that access to a given resource will be denied unless a user can provide a specific business reason or need for access to the information resource. The converse of this access philosophy—which is not widely accepted—is the accept-all philosophy, under which everyone is allowed access unless someone can provide areas for denying access.

FIREWALL GENERAL FEATURES

Firewalls separate networks from one another and screen the traffic between them. See **figure 3.7**.

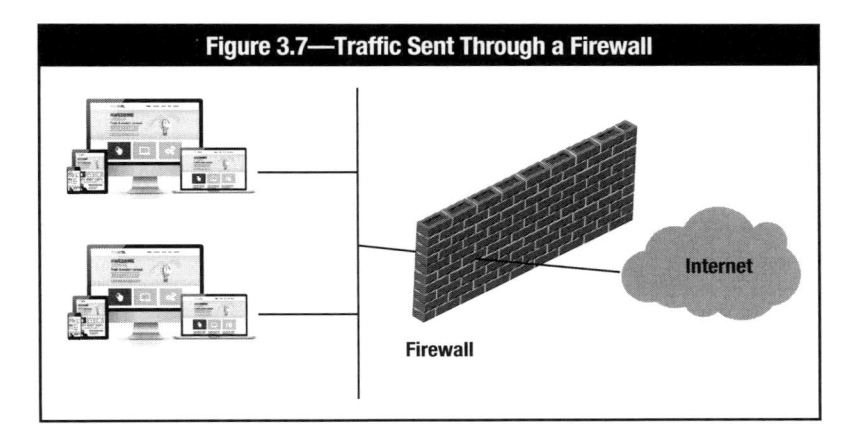

Figure 3.7—Traffic Sent Through a Firewall

Along with other types of security (e.g., intrusion detection systems [IDS]/intrusion prevention systems [IPS]), firewalls control the most vulnerable point between a corporate network and the Internet, and they can be as simple or complex as the corporate information security policy demands.

There are many different types of firewalls, but most of them enable organizations to:
- Block access to particular sites on the Internet.
- Limit traffic on an organization's public services segment to relevant addresses and ports.
- Prevent certain users from accessing certain servers or services.
- Monitor and record communications between an internal and an external network in order to investigate network penetrations or detect internal subversion.
- Encrypt packets that are sent between different physical locations within an organization by creating a VPN over the Internet (e.g., IP security [IPSec], secure VPN tunnels). The capabilities of some firewalls can be extended so they can also provide for protection against viruses and attacks directed to exploit known operating system vulnerabilities.

Section 3: Security Architecture Principles

NETWORK FIREWALL TYPES

Generally, the types of network firewalls available today fall into the following categories:
- Packet filtering
- Application firewall systems
- Stateful inspection
- Next generation firewall (NGFW)

Each type of firewall is discussed in the following sections. A summary is provided in **figure 3.8**.

	Figure 3.8—Firewall Types
First Generation	A simple packet-filtering router that examines individual packets and enforces rules based on addresses, protocols and ports.
Second Generation	Keeps track of all connections in a state table. This allows it to enforce rules based on packets in the context of the communications session.
Third Generation	Operates at layer seven (the application layer) and is able to examine the actual protocol being used for communications, such as Hypertext Transfer Protocol (HTTP). These firewalls are much more sensitive to suspicious activity related to the content of the message itself, not just the address information.
Next Generation	Sometimes called deep packet inspection—is an enhancement to third generation firewalls and brings in the functionality of an intrusion prevention system (IPS) and will often inspect Secure Sockets Layer (SSL) or Secure Shell (SSH) connections.

Source: ISACA, *CRISC Review Manual 6th Edition*, USA, 2015

PACKET FILTERING FIREWALLS

First generation firewalls were packet filtering-based firewalls deployed between the private network and the Internet. In packet filtering, a screening router examines the header of every packet of data traveling between the Internet and the corporate network. Packet headers contain information, including the IP address of the sender and receiver, along with the port numbers (application or service) authorized to use the information transmitted. Based on that information, the router knows what kind of Internet service (e.g., web-based service or File Transfer Protocol [FTP]) is being used to send the data as well as the identities of the sender and receiver of the data. Then, the router can prevent certain packets from being sent between the Internet and the corporate network. For example, the router could block any traffic to and from suspicious destinations. See **figure 3.9**.

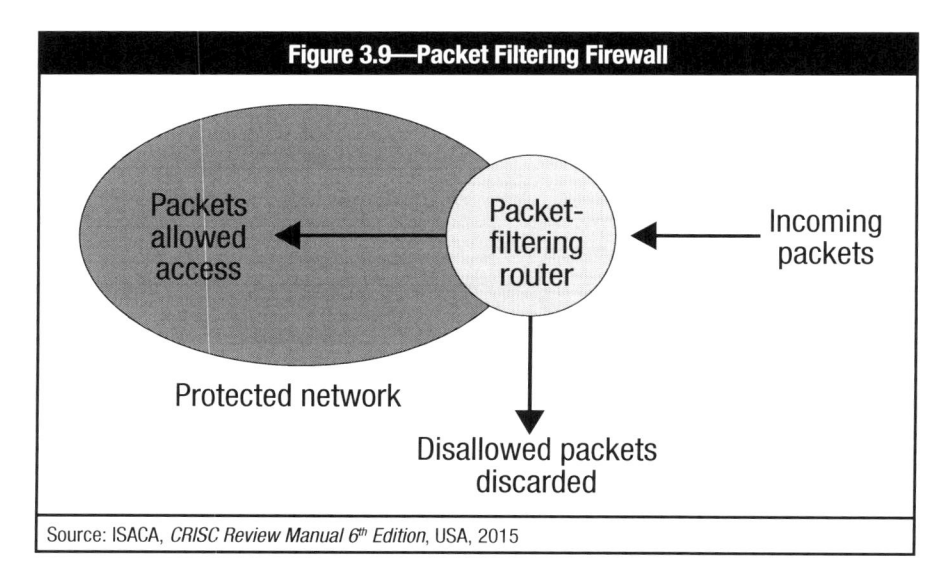

Figure 3.9—Packet Filtering Firewall

Packets allowed access ← Packet-filtering router ← Incoming packets

Protected network

Disallowed packets discarded

Source: ISACA, *CRISC Review Manual 6th Edition*, USA, 2015

Because the direct exchange of packets is permitted between outside systems and inside systems, the potential for an attack is determined by the total number of hosts and services to which the packet filtering router permits traffic. Packet filtering firewalls are, therefore, best suited for smaller networks. Organizations with many routers may face difficulties in designing, coding and maintaining the rule base.

Because their filtering rules are performed at the network layer, packet filtering firewalls are generally stable and simple. This simplicity has both advantages and disadvantages, as shown in **figure 3.10**.

Figure 3.10—Packet Filtering Firewalls	
Advantages	**Disadvantages**
Simplicity of one network "choke point"	Vulnerable to attacks from improperly configured filters
Minimal impact on network performance	Vulnerable to attacks tunneled over permitted services
Inexpensive or free	All private network systems vulnerable when a single packet filtering router is compromised

In light of these advantages and disadvantages, packet filtering is most effective when implemented with basic security and monitoring in mind.

Some of the more common attacks against packet filtering firewalls are:
- **IP spoofing**—In this type of attack, the attacker fakes the IP address of either an internal network host or a trusted network host. This enables the packet being sent to pass the rule base of the firewall and penetrate the system perimeter. If the spoofing uses an internal IP address, the firewall can be configured to drop the packet on the basis of packet flow direction analysis. However, attackers with access to a secure or trusted external IP address can spoof on that address, leaving the firewall architecture defenseless.
- **Source routing specification**—This type of attack centers around the routing that an IP packet must take when it traverses the Internet from the source host to the destination host. In this process, it is possible to define the route so it bypasses the firewall. However, the attacker must know the IP address, subnet mask and default gateway settings to accomplish this. A clear defense against this attack is to examine each packet and drop packets that have source routing specification enabled. Note that this countermeasure will not be effective if the topology permits a route that skips the choke point.
- **Miniature fragment attack**—Using this method, an attacker fragments the IP packet into smaller ones and pushes it through the firewall. This is done with the hope that only the first sequence of fragmented packets will be examined, allowing the others to pass without review. This is possible only if the default setting is to pass residual packets. Miniature fragment attacks can be countered by configuring the firewall to drop all packets where IP fragmentation is enabled.

APPLICATION FIREWALL SYSTEMS

Packet filtering routers allow the direct flow of packets between internal and external systems. The primary risk of allowing packet exchange between internal and external systems is that the host applications residing on the protected network's systems must be secure against any threat posed by the allowed packets.

In contrast to packet filtering routers, application- and circuit-level gateways allow information to flow between systems but do not allow the direct exchange of packets. Therefore, application firewall systems provide greater protection capabilities than packet filtering routers.

The two types of application firewall systems sit atop hardened (i.e., tightly secured) operating systems such as Windows® and UNIX®. They work at the application level of the OSI model.

The two types of application firewall systems are:
- **Application-level gateways**—Application-level gateways are systems that analyze packets through a set of proxies—one for each service (e.g., Hypertext Transmission Protocol [HTTP] proxy for web traffic, FTP proxy). The implementation of multiple proxies, however, impacts network performance. When network performance is a concern, a circuit-level gateway may be a better choice.

- **Circuit-level gateways**—Commercially, circuit-level gateways are quite rare. Because they use one proxy server for all services, they are more efficient and also operate at the application level. There, TCP and UDP sessions are validated, typically through a single, general-purpose proxy before opening a connection. This differs from application-level gateways, which require a special proxy for each application-level service.

Both application firewall systems employ the concept of bastion hosting in that they handle all incoming requests from the Internet to the corporate network, such as FTP or web requests. Bastion hosts are heavily fortified against attack. When there is only one host handling incoming requests, it is easier to maintain security and track attacks. In the event of a break-in, only the firewall system is compromised, not the entire network.

This way, none of the computers or hosts on the corporate network can be contacted directly for requests from the Internet, providing an effective level or layer of security.

Additionally, application-based firewall systems are set up as proxy servers to act on the behalf of someone inside an organization's private network. Rather than relying on a generic packet-filtering tool to manage the flow of Internet services through the firewall, a special-purpose code called a proxy server is incorporated into the firewall system. For example, when someone inside the corporate network wants to access a server on the Internet, a request from the computer is sent to the proxy server. The proxy server contacts the Internet server, and the proxy server then sends the information from the Internet server to the computer inside the corporate network. By acting as a go-between, proxy servers can maintain security by examining the program code of a given service (e.g., FTP, Telnet). It then modifies and secures it to eliminate known vulnerabilities. The proxy server can also log all traffic between the Internet and the network.

One feature available on both types of firewall systems is the network address translation (NAT) capability. This capability takes private internal network addresses, which are unusable on the Internet, and maps them to a table of public IP addresses assigned to the organization, which can be used across the Internet.

Application firewalls have advantages and disadvantages, as shown in **figure 3.11**.

Figure 3.11—Application Firewalls

Advantages	Disadvantages
Provide security for commonly used protocols	Reduced performance and scalability as Internet usage grows
Generally hide the network from outside untrusted networks	
Ability to protect the entire network by limiting break-ins to the firewall itself	
Ability to examine and secure program code	

STATEFUL INSPECTION FIREWALLS

A stateful inspection firewall, also referred to as dynamic packet filtering, tracks the destination IP address of each packet that leaves the organization's internal network. Whenever a response to a packet is received, its record is referenced to ascertain whether the incoming message was made in response to a request that the organization sent out. This is done by mapping the source IP address of an incoming packet with the list of destination IP addresses that is maintained and updated. This approach prevents any attack initiated and originated by an outsider.

In contrast to application firewalls, stateful inspection firewalls provide control over the flow of IP traffic. They do this by matching information contained in the headers of connection-oriented or connectionless IP packets at the transport layer against a set of rules authorized by the organization. Consequently, they have advantages and disadvantages, as shown in **figure 3.12**.

Figure 3.12—Stateful Inspection Firewalls	
Advantages	**Disadvantages**
Provide greater control over the flow of IP traffic	Complex to administer
Greater efficiency in comparison to CPU-intensive, full-time application firewall systems	

STATELESS VS. STATEFUL

Stateless filtering does not keep the state of ongoing TCP connection sessions. In other words, it has no memory of what source port numbers the sessions' client selected. Stateful firewalls keep track of TCP connections. The firewall keeps an entry in a cache for each open TCP connection. Stateless firewalls perform more quickly than stateful firewalls, but they are not as sophisticated.

As UDP traffic is stateless, applications that require UDP to operate from the Internet into a corporate network should be used sparingly and/or alternate controls implemented (e.g., network segregation or use of application aware firewalls, NGFWs and application-aware IDS/IPS).

EXAMPLES OF FIREWALL IMPLEMENTATIONS

Firewall implementations can take advantage of the functionality available in a variety of firewall designs to provide a robust layered approach in protecting an organization's information assets. Commonly used implementations available today include:

- **Screened-host firewall**—Using a packet filtering router and a bastion host, this approach implements basic network layer security (packet filtering) and application server security (proxy services). An intruder in this configuration must penetrate two separate systems before the security of the private network can be compromised. This firewall system is configured with the bastion host connected to the private network with a packet filtering router between the Internet and the bastion host. Router filtering rules allow inbound traffic to access only the bastion host, which blocks access to internal systems. Because the inside hosts reside on the same network as the bastion host, the security policy of the organization determines whether inside systems are permitted direct access to the Internet, or whether they are required to use the proxy services on the bastion host.
- **Dual-homed firewall**—This is a firewall system that has two or more network interfaces, each of which is connected to a different network. A dual-homed firewall usually acts to block or filter some or all of the traffic trying to pass between the networks. A dual-homed firewall system is a more restrictive form of a screened-host firewall system in which a dual-homed bastion host is configured with one interface established for information servers and another for private network host computers.
- **Demilitarized zone (DMZ) or screened-subnet firewall**—As shown in **figure 3.13**, this is a small, isolated network for an organization's public servers, bastion host information servers and modem pools. The DMZ connects the untrusted network to the trusted network, but it exists in its own independent space to limit access and availability of resources. As a result, external systems can access only the bastion host and possibly information servers in the DMZ. The inside router manages access to the private network, accepting only traffic originating from the bastion host. The filtering rules on the outside router require the use of proxy services by accepting only outbound traffic on the bastion host. The key benefits of this system are that an intruder must penetrate three separate devices, private network addresses are not disclosed to the Internet and internal systems do not have direct access to the Internet.

Section 3: Security Architecture Principles

Figure 3.13—The Demilitarized Zone

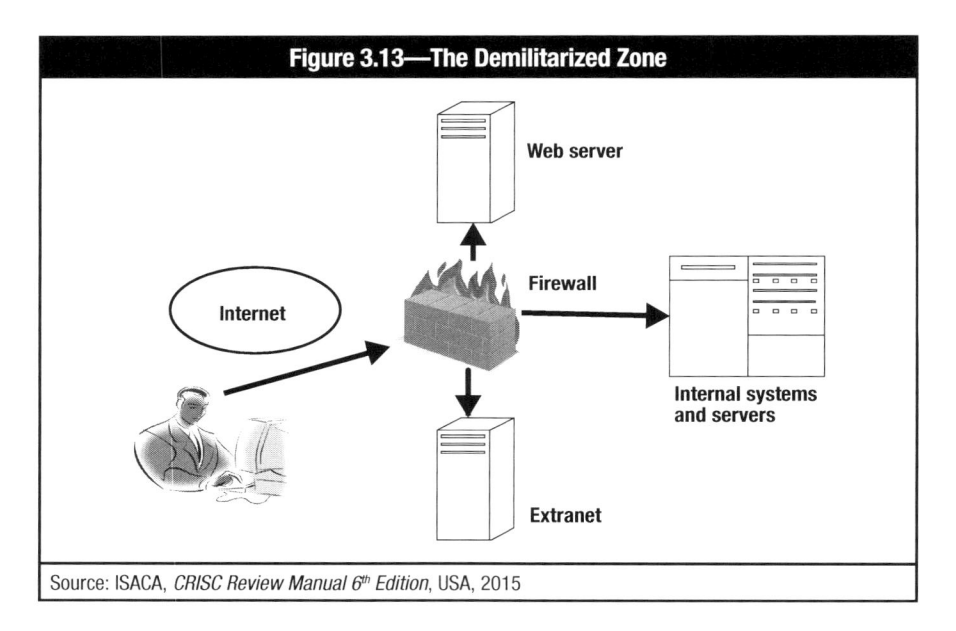

Source: ISACA, *CRISC Review Manual 6th Edition*, USA, 2015

FIREWALL ISSUES

Problems faced by organizations that have implemented firewalls include the following:
- **Configuration errors**—Misconfigured firewalls may allow unknown and dangerous services to pass through freely.
- **Monitoring demands**—It is necessary to apply and review log settings appropriately, but monitoring activities may not always occur on a regular basis.
- **Policy maintenance**—Firewall policies may not be maintained regularly.
- **Vulnerability to application- and input-based attacks**—Most firewalls operate at the network layer; therefore, they do not stop any application-based or input-based attacks, such as SQL injection and buffer-overflow attacks. Newer generation firewalls are able to inspect traffic at the application layer and stop some of these attacks.

FIREWALL PLATFORMS

Firewalls may be implemented using hardware, software or virtual platforms. Implementing hardware will provide performance with minimal system overhead. Although hardware-based firewall platforms are faster, they are not as flexible or scalable as software-based firewalls. Software-based firewalls are generally slower with significant systems overhead. However, they are flexible with additional services; for example, they may include content and virus checking before traffic is passed to users.

It is generally better to use appliances, rather than normal servers, for the firewall. An appliance is a device with all software and configurations pre-setup on a physical server that is plugged in between two networks. Appliances are normally installed with hardened operating systems. When server-based firewalls are used, operating systems in servers are often vulnerable to attacks. When attacks on operating systems succeed, the firewall can be compromised. In general, appliance-type firewalls are significantly faster to set up and recover.

NEXT GENERATION FIREWALLS[37, 38, 39, 40]

NGFWs are firewalls aimed at addressing two key limitations of earlier variants: 1) the inability to inspect packet payload and 2) the inability to distinguish between types of web traffic. An NGFW is an adaptive network security system capable of detecting and blocking sophisticated attacks. NGFWs typically perform traditional functions such as packet filtering, stateful inspection and network address translation (NAT), but introduce application awareness, incorporate deep packet inspection (DPI) technology and offer varying degrees of integrated threat protection, such as data loss prevention (DLP), intrusion prevention system (IPS), secure sockets layer (SSL)/secure shell (SSH) inspection and web filtering.

Application awareness is "the capacity of a system to maintain information about connected applications to optimize their operation and that of any subsystems that they run or control."[41] This is important because discriminating between legitimate and malicious traffic has become increasingly difficult amid the upsurge in web-based services. The ability of an NGFW to differentiate between types of web traffic such as an authorized business web application and a streaming media site aids enforcement of corporate policies—regardless of port or protocol—and similarly offers insight to user activities and behavior.

DPI allows for payload interrogation against signatures for known exploits, malware, etc. DPI affords a great deal of information about your traffic, which aids in determination of normal traffic making anomaly detection more effective, especially in more complex networks.

Depending on your organization, you may be asked to review, recommend or specify vendor solutions. Bear in mind that while many next-generation solutions advertise similar functions, how they do so is often decided by their interpretation of concepts and implementation of proprietary technology. As sophisticated as NGFWs may be today, it should not be your only line of defense.

WEB APPLICATION FIREWALLS[42]

A web application firewall (WAF) is a server plug-in, appliance or additional filter that can be used to apply rules to a specific web application (usually to an HTTP conversation). It operates at the higher levels of the OSI model, generally level 7, while network firewalls operate at level 3 and/or level 4. It can be customized to identify and block many types of attacks, such as cross-site scripting (XSS) and Structured Query Language (SQL) injection. Customization of the WAF rules requires a lot of work and effort. When changes to the application are made, the WAF rules need to be changed as well.

---

[37] Ohlhorst, Frank, "Next-Generation Firewalls 101," Network Computing, 1 March 2013, www.networkcomputing.com/careers-and-certifications/next-generation-firewalls-101/a/d-id/1234097
[38] Miller, Lawrence C.; *Next-Generation Firewalls for Dummies*, Wiley Publishing, Inc., USA, 2011
[39] Rouse, Margaret, "Next-generation firewall," January 2014, TechTarget, SearchSecurity, http://searchsecurity.techtarget.com/definition/next-generation-firewall-NGFW
[40] My Digital Shield, "Firewalls Don't Cut It Anymore – Why You Need Next Generation Firewalls," 28 August 2014, www.mydigitalshield.com/firewalls-dont-cut-anymore-need-next-generation-firewalls
[41] Wigmore, Ivy; "Application Awareness," January 2013, TechTarget, http://whatis.techtarget.com/definition/application-awareness
[42] Open Web Application Security Project (OWASP), *Web Application Firewall*, www.owasp.org/index.php/Web_Application_Firewall

Section 3: Security Architecture Principles

TOPIC 5—ISOLATION AND SEGMENTATION

VIRTUAL LOCAL AREA NETWORKS

A common technique for implementing network security is to segment an organization's network so that each segment can be separately controlled, monitored and protected. **Virtual local area networks (VLANs)** are groups of devices on one or more logically segmented LAN, usually without additional encryption used.

A VLAN is set up by configuring ports on a switch, so devices attached to these ports may communicate as if they were attached to the same physical network segment, although the devices are actually located on different LAN segments. Segmenting network traffic in this way enables an organization to keep different types of data separate from one another.

A VLAN is based on logical rather than physical connections, and thus, it allows great flexibility. This flexibility enables administrators to segment network resources for optimal performance by restricting users' access of network resources to the necessary individuals only. In Layer 4 switching (transport layer), some application information is taken into account along with Layer 3 addresses. For IP, this information includes the port numbers from protocols such as UDP and TCP. These devices, unlike Layer 3 switches, are more resource intensive because they have to store application-based protocol information. Only address information is stored at the Layer 2 and Layer 3 levels.

SECURITY ZONES AND DEMILITARIZED ZONES

By creating separate zones, controls can be applied at a more granular level based on the systems, information and applications in each area. Separate zones can create defense in depth where additional layers of authentication, access control and monitoring can take place. Isolation and segmentation is shown in **figure 3.14**.

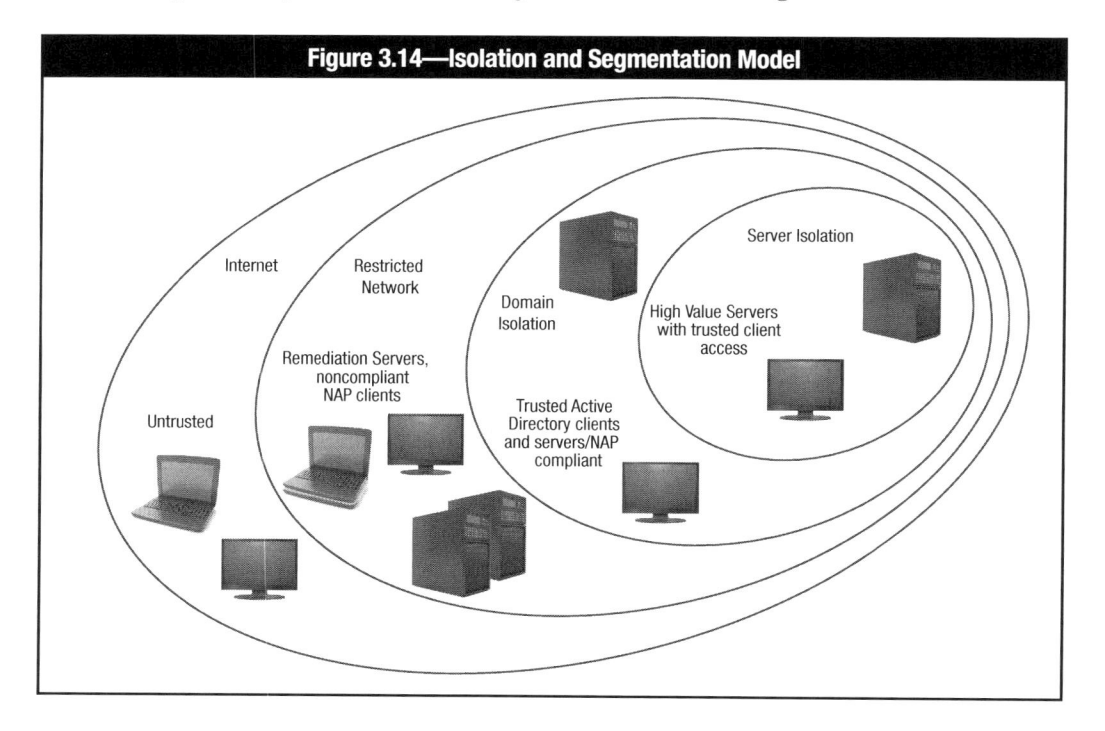

Figure 3.14—Isolation and Segmentation Model

Most organizations separate their internal systems from the Internet using a firewall. However, some systems and services, such as web servers, need to be available outside the internal network. This can be accomplished with a network segment called a **demilitarized zone (DMZ)**, which places limited systems, applications and data in a public-facing segment. Servers located in a DMZ minimize the exposure to attacks.

The DMZ functions as a small, isolated network for an organization's public servers, virtual private network (VPN) termination and modem pools. Typically, DMZs are configured to limit access from the Internet and the organization's private network. Incoming traffic access is restricted into the DMZ network by the outside router and firewall, protecting the organization against certain attacks by limiting the services available for use. Consequently, external systems can access only the systems in the DMZ.

Page intentionally left blank

TOPIC 6—LOGGING, MONITORING AND DETECTION

Monitoring, detection and logging are integral parts of cybersecurity. With potential for attacks and data loss on both sides, it is necessary to monitor data and information flowing into and out of an organization. As this topic will illustrate, there are a number of methods and tools an organization can use to detect and log potential problems. Most of these methods revolve around the central concepts of ingress, egress and data loss prevention.

LOGGING[43]

A log is a record of events that occur within the systems and networks of an organization. Logs are one of the most valuable tools to monitor controls and detect risk, but they are often underutilized. A log should contain a record of all important events that occur on a system, and should include:
- Time of the event
- Changes to permissions
- System startup or shutdown
- Login or logout
- Changes to data
- Errors or violations
- Job failures

Log reviews can identify risk-relevant events such as compliance violations, suspicious behavior, errors, probes or scans, and abnormal activity. A failure to review the logs can result in the organization not being aware of an ongoing attack. Logs may be required to comply with legislation and regulatory compliance and should also be preserved for forensic analysis if needed at a later time. The cybersecurity professional may find it useful to employ analysis tools when reviewing logs in order to filter pertinent data.

Ensuring proper segregation of duties (SoD) and time synchronization is particularly important when it comes to log files. The ability to change system configurations should be segregated from the ability to review, modify or delete logs in order to ensure that the organization can exercise proper oversight of administrative functions. Without time synchronization, it is extremely difficult to correlate information from different logs (server, router, firewall) to analyze the occurrence of an event.

Common challenges relating to the effective use of logs include:
- Having too many data
- Difficulty in searching for relevant information
- Improper configuration (e.g., may not be enabled or contain appropriate data)
- Modification or deletion of data before they are read (e.g., too little storage space)

SECURITY INFORMATION AND EVENT MANAGEMENT

To prepare for and identify an incident, organizations use a myriad of security tools, such as vulnerability assessments, firewalls and IDSs, that collect a high volume of data. However, security teams have to analyze and interpret this overwhelming amount of data, referred to as log data overload. An emerging solution to this problem is security event management (SEM). SEM systems automatically aggregate and correlate security event log data across multiple security devices. This allows security analysts to focus on a manageable list of critical events.

Security incidents are often made up of a series of events that occur throughout a network. By correlating data, the SEM can take many isolated events and combine them to create one single relevant security incident. These systems use either rule-based or statistical correlation. Rule-based correlations create situation-specific rules that establish a pattern of events. Statistical correlation uses algorithms to calculate threat levels incurred by relevant events on various IT assets.

---

[43] ISACA, *CRISC Review Manual 6th Edition*, USA, 2015

Section 3: Security Architecture Principles

There are a variety of SEM solutions available that provide real-time monitoring, correlation of events, notifications and console views. In addition, security information and event management (SIEM) systems take the SEM capabilities and combine them with the historical analysis and reporting features of security information management (SIM) systems.

Information security teams should periodically analyze the trends found from SEM or SIEM systems, such as attempted attack types or most frequently targeted resources. This allows the organization to investigate incidents as well as allocate appropriate resources to prevent future incidents.

One organization that often uses SIEM for monitoring and detection is a security operations center (SOC). A SOC consists of an organized team created to improve the security posture of an organization and to respond to cybersecurity incidents.[44]

INGRESS, EGRESS AND DATA LOSS PREVENTION

There are two types of attack vectors: ingress and egress (also known as data exfiltration). **Ingress** refers to network communications coming in, while **egress** refers to network communications going out. While most attack analysis concentrates on the ingress or intrusion into systems, if the adversary's goal is theft of information or data, then it is important to consider the vector or path used to remove the data from the owner's systems and networks. Data loss prevention (DLP) software is helpful in this regard. A successful data loss prevention program helps an organization protect its information and prevent the exfiltration of sensitive data.

Strong DLP solutions cover three primary states of information. **Data at rest** refers to stored data. DLP solutions must be able to log where various file types are stored. Crawler applications then explore the information on these files searching for sensitive data like social security or credit card information. These crawlers determine whether the storage location follows predefined rules.

Data in transit refers to data traveling through the network. Deep packet inspection (DPI) is used to analyze the data for sensitive content. DLP solutions can alert management and even block, quarantine or encrypt controlled information based on controls.

Finally, good DLP solutions manage **data in use**, which is data movement at the user workstation level. This includes sending information to printers, thumb drives or even the copy-and-paste clipboard. DLP solutions use agent software to set rules for data use. All three information types, data at rest, data in motion and data in use, must be addressed to create a full DLP solution.

ANTIVIRUS AND ANTI-MALWARE

Malicious software is one of the most common attack vectors used by adversaries to compromise systems. Therefore, controls are required for its detection and prevention.

Historically, anti-malware controls, often referred to as virus checkers, were host-based applications that scanned incoming traffic such as email and looked for patterns (signatures) that identified known problems. While this can be effective for known threats, it cannot detect malicious code that has yet to be identified.

Heuristic-based methods of detecting unknown malware use specific techniques to identify common malicious code behaviors and flag them as suspicious.

Anti-malware can be controlled through many different mechanisms, including:
- Restriction of outbound traffic to prevent malware from exfiltrating data or communicating with control systems used by the adversary
- Policies and awareness that train users to avoid opening suspect emails or attachments and to recognize Uniform resource locators (URLs) that may introduce malicious code
- Multiple layers of anti-malware software using a combination of signature identification and heuristic analysis to identify possible malicious code

---

[44] Paganini, Pierluigi, "What is a SOC (Security Operations Centers)?," Security Affairs, 24 May 2016, *http://securityaffairs.co/wordpress/47631/breaking-news/soc-security-operations-center.html*

INTRUSION DETECTION SYSTEMS[45]

Another element to securing networks that complements firewall implementation is an IDS. An IDS works in conjunction with routers and firewalls by monitoring network usage anomalies. It protects a company's IS resources from external as well as internal misuse. An IDS operates continuously on the system, running in the background and notifying administrators when it detects a perceived threat. Broad categories of IDSs include:

- **Network-based IDSs**—These identify attacks within the monitored network and issue a warning to the operator. If a network-based IDS is placed between the Internet and the firewall, it will detect all the attack attempts, regardless of whether they enter the firewall. If the IDS is placed between a firewall and the corporate network, it will detect those attacks that enter the firewall (i.e., it will detect intruders). The IDS is not a substitute for a firewall, but rather it complements the function of a firewall.
- **Host-based IDSs**—These are configured for a specific environment and will monitor various internal resources of the operating system to warn of a possible attack. They can detect the modification or execution of files and issue a warning when an attempt is made to run a privileged command.

Components of an IDS are:
- Sensors responsible for collecting data in the form of network packets, log files, system call traces, etc.
- Analyzers that receive input from sensors and determine intrusive activity
- An administration console

Types of IDSs include:
- **Signature-based**—These IDS systems protect against detected intrusion patterns. The intrusive patterns they can identify are stored in the form of signatures.
- **Statistical-based**—These systems need a comprehensive definition of the known and expected behavior of systems.
- **Neural networks**—An IDS with this feature monitors the general patterns of activity and traffic on the network and creates a database. It is similar to the statistical model but with added self-learning functionality.

Signature-based IDSs are not able to detect all types of intrusions due to the limitations of their detection rules. On the other hand, statistical-based systems may report many events outside of the defined normal activity that are still normal activities on the network. A combination of signature- and statistical-based models provides better protection.

IDS Features
The features available in an IDS include:
- Intrusion detection
- Ability to gather evidence on intrusive activity
- Automated response (e.g., termination of connection, alarm messaging)
- Security policy
- Interface with system tools
- Security policy management

IDS Limitations
An IDS cannot help with the following weaknesses:
- Weaknesses in the policy definition (see Policy section)
- Application-level (programming) vulnerabilities
- Back doors into applications
- Weaknesses in identification and authentication schemes

---

[45] ISACA, *CISA Review Manual 26th Edition*, USA, 2015

IDS Policy
An IDS policy should establish the action to be taken by security personnel in the event that an intruder is detected.

Actions include:
- **Terminate the access**—If there is a significant risk to the organization's data or systems, immediate termination is the usual procedure.
- **Trace the access**—If the risk to the data is low, the activity is not immediately threatening, or analysis of the entry point and attack method is desirable, the IDS can be used to trace the origin of the intrusion. This can be used to determine and correct any system weaknesses and to collect evidence of the attack that may be used in a subsequent court action.

In either case, the action required should be determined by management in advance and incorporated into a procedure. This will save time when an intrusion is detected, which may impact the possible data loss.

INTRUSION PREVENTION SYSTEMS
An IPS is a system designed to not only detect attacks, but also to prevent the intended victim hosts from being affected by the attacks. It complements firewall, antivirus and antispyware tools to provide a more complete protection from emerging threats.

IPS technology is commonly placed at the perimeter of the enterprise network—at all ingress/egress points—to examine network traffic flows and prevent zero-day attacks, such as worms or viruses. The detection methods used by an IPS are anomaly-based rules and signature-based inspection of network packets. A well-managed IPS solution helps to ensure that threats are dropped at the network perimeter; an IDS provides visibility and confirmation of inside activity that is at critical network nodes.

The biggest advantage of an IPS is that it can help block an attack when it occurs; rather than simply sending an alert, it actively helps to block malicious and unwanted traffic.

However, as with an IDS, the IPS must be properly configured and tuned to be effective. Threshold settings that are too high or low will lead to limited effectiveness of the IPS. Additionally, some concerns have been raised that IPSs may in themselves constitute a threat, because a clever attacker could send commands to a large number of hosts protected by an IPS in order to cause them to become dysfunctional. Such a situation could have a potentially catastrophic outcome in today's typical corporate computing environment where continuity of service is critical. In addition, IPSs can generate false positives that can create serious problems if automated responses are used.

TOPIC 7—ENCRYPTION FUNDAMENTALS, TECHNIQUES AND APPLICATIONS

Encryption is the process of converting a plaintext message into a secure-coded form of text, called ciphertext. The ciphertext cannot be understood without converting back, via decryption—the reverse process—to plaintext. This is done via a mathematical function and a special encryption/decryption password called the key. In many countries, encryption is subject to governmental laws and regulations that limit the key size or define what may not be encrypted.

Encryption is part of a broader science of secret languages called cryptography, which is generally used to:
- Protect information stored on computers from unauthorized viewing and manipulation
- Protect data in transit over networks from unauthorized interception and manipulation
- Deter and detect accidental or intentional alterations of data
- Verify authenticity of a transaction or document

Encryption is limited in that it cannot prevent the loss of data. It is possible to compromise encryption programs if encryption keys are not protected adequately. Therefore, encryption should be regarded as an essential, but incomplete, form of access control that should be incorporated into an organization's overall computer security program.

KEY ELEMENTS OF CRYPTOGRAPHIC SYSTEMS

Key elements of cryptographic systems include:
- **Encryption algorithm**—Mathematically based function or calculation that encrypts or decrypts data.
- **Encryption key**—Piece of information similar to a password that makes the encryption or decryption process unique. A user needs the correct key to access or decipher a message, as the wrong key converts the message into an unreadable form.
- **Key length**—Predetermined length for the key. The longer the key, the more difficult it is to compromise in a brute force attack where all possible key combinations are tried.

Effective cryptographic systems depend upon a variety of factors including:
- Algorithm strength
- Secrecy and difficulty of compromising a key
- Nonexistence of back doors by which an encrypted file can be decrypted without knowing the key
- Inability to decrypt parts of a ciphertext message and prevent known plaintext attacks
- Properties of the plaintext known by a perpetrator

KEY SYSTEMS

There are two types of cryptographic systems:
- **Symmetric key systems**—These use single, secret, bidirectional keys that encrypt and decrypt.
- **Asymmetric key systems**—These use pairs of unidirectional, complementary keys that only encrypt or decrypt. Typically, one of these keys is secret, and the other is publicly known.

Public key systems are asymmetric cryptographic systems. Most encrypted transactions over the Internet use a combination of private/public keys, secret keys, hash functions (fixed values derived mathematically from a text message) and digital certificates (that prove ownership of a public encryption key) to achieve confidentiality, message integrity, authentication and nonrepudiation by either sender or recipient (also known as a public key infrastructure [PKI]). Essentially, keys and hash values are used to transform a string of characters into a shorter or fixed-length value or key that represents the original string. This encryption process allows data to be stored and transported with reduced exposure so data remain secure as they move across the Internet or other networks.

Section 3: Security Architecture Principles

SYMMETRIC (PRIVATE) KEY ENCRYPTION

Symmetric key cryptographic systems are based on a symmetric encryption algorithm, which uses a secret key to encrypt the plaintext to the ciphertext and the same key to decrypt the ciphertext to the corresponding plaintext. In this case, the key is said to be symmetric because the encryption key is the same as the decryption key. An example of symmetric cryptography is shown in **figure 3.15**.

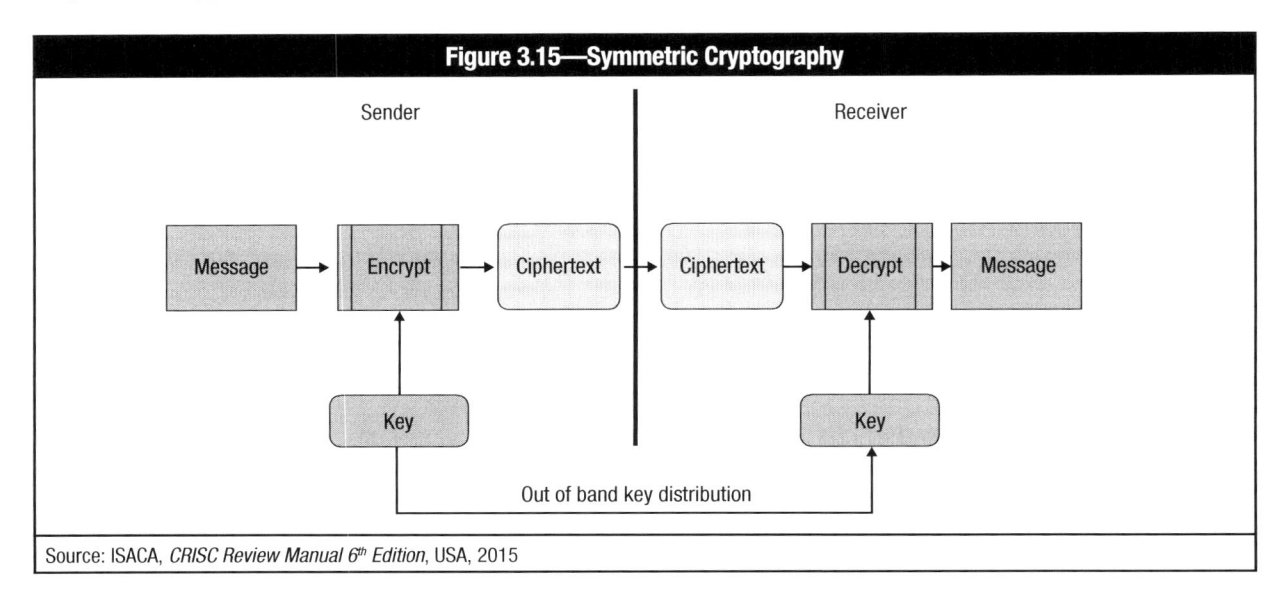

Figure 3.15—Symmetric Cryptography

Source: ISACA, *CRISC Review Manual 6th Edition*, USA, 2015

The most common symmetric key cryptographic system used to be the Data Encryption Standard (DES). DES is based on a public algorithm approved by the National Institute of Standards and Technology (NIST) and employs keys of 56 bits (plus 8 bits used for parity checking). The bits in the plaintext are processed one 64-bit block at a time and, as such, DES belongs to the category of block-ciphers (as opposed to stream-ciphers, which encode one bit at a time).[46]

DES has been withdrawn by NIST because it is no longer considered a strong cryptographic solution because its entire key space can be brute forced by a moderately large computer system within a relatively short period of time. Extensions of DES (Triple DES or 3DES) were proposed to extend the DES standard while retaining backward compatibility (it applies the DES cipher algorithm three times to each data block).[47]

In 2001, NIST replaced DES with the Advanced Encryption Standard (AES), a public algorithm that supports keys from 128 bits to 256 bits in size. Another commonly used symmetric key algorithm, although considered breakable and insecure, is RC4, a stream-cipher often used in SSL/TLS protocol sessions.[48]

There are two main advantages to symmetric key cryptosystems such as DES or AES:
- The user only has to know one key for both encryption and decryption.
- Symmetric key cryptosystems are generally less complicated and, therefore, use up less processing power than asymmetric techniques. They are ideally suited for bulk data encryption.

The disadvantages of this approach include:
- Difficulty distributing keys—Getting the keys into the hands of those with whom you want to exchange data can be a challenge, particularly in ecommerce environments where customers are unknown, untrusted entities.
- Limitations of shared secret—A symmetric key cannot be used to sign electronic documents or messages due to the fact that the mechanism is based on a shared secret.

One form of advanced encryption algorithm is known as Triple DES or 3DES. Triple DES provides a relatively simple method of increasing the key size of DES to protect information without the need to design a completely new block cipher algorithm.

---

[46] ISACA, *CISA Review Manual 26th Edition*, USA, 2015
[47] *Ibid.*
[48] *Ibid.*

ASYMMETRIC (PRIVATE) KEY ENCRYPTION

Public key cryptographic systems developed for key distribution solve the problem of getting single symmetric keys into the hands of two people who do not know each other but who want to exchange information securely. Based on an asymmetric encryption process, two keys work together as a pair. One key is used to encrypt data; the other is used to decrypt data. Either key can be used to encrypt or decrypt, but once the key has been used to encrypt data, only its partner can be used to decrypt the data. The key that was used to encrypt the data cannot be used to decrypt it. Thus, the keys are asymmetric in that they are inversely related to each other.

Asymmetric keys are often used for short messages such as encrypting DES symmetric keys or creating digital signatures. If asymmetric keys were used to encrypt bulk data (long messages), the process would be very slow; this is the reason they are used to encrypt short messages such as digests or signatures.

With asymmetric encryption, one key—the secret or private key—is known only to one person. The other key—the public key—is known by many people. In other words, a message that has been sent encrypted by the secret (private) key of the sender can be deciphered by anyone with the corresponding public key. In this way, if the public key deciphers the message satisfactorily, one can be sure of the origin of the message because only the sender (owner of the correspondent private key) could have encrypted the message. This forms the basis of authentication and nonrepudiation, as the sender cannot later claim that he or she did not generate the message.

A message that has been sent encrypted using the public key of the receiver can be generated by anyone, but can only be read by the receiver. This is one basis of confidentiality. In theory, a message that has been encrypted twice, first by the sender's secret key and second by the receiver's public key, achieves both authentication and confidentiality objectives, but it is not commonly used because it could generate performance issues.

One disadvantage to using asymmetric algorithms is they are computationally intensive and slow relative to symmetric algorithms. For that reason, asymmetric cryptography is typically used only to encrypt short messages. In fact, the most common use of asymmetric algorithms is to distribute symmetric keys that can then be used by the participants for fast, secure communication, as seen in **figure 3.16**.[49]

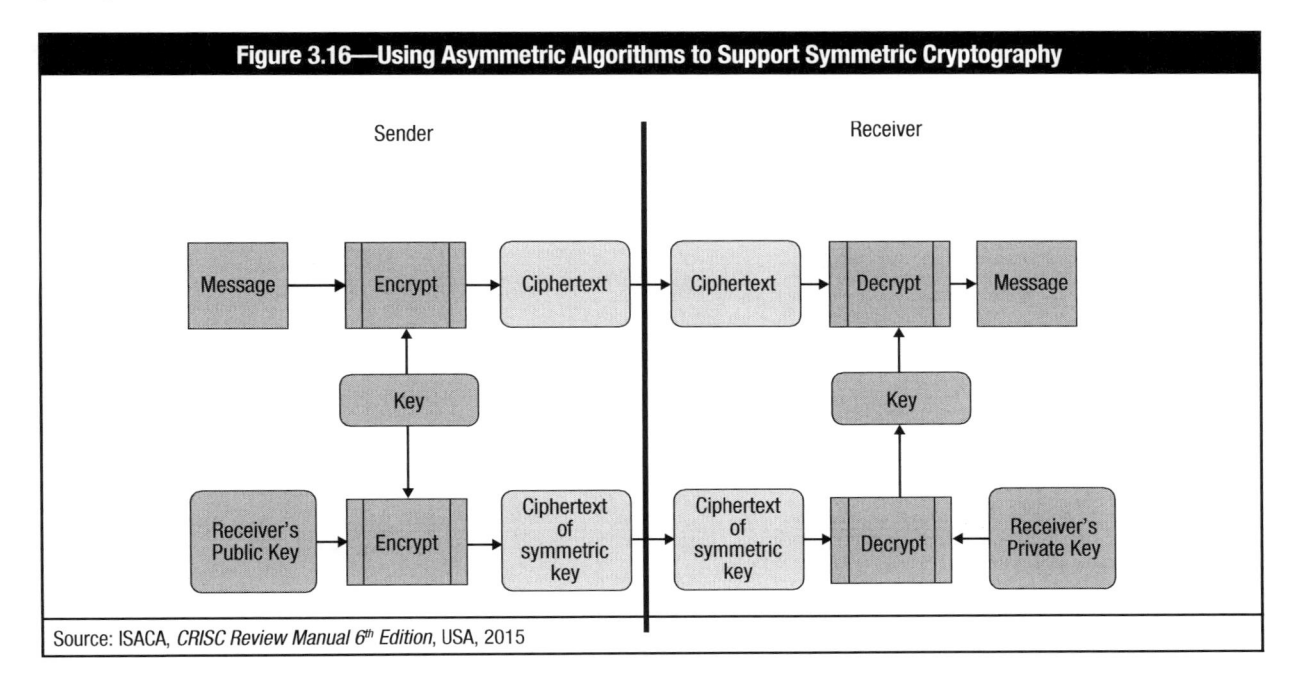

Figure 3.16—Using Asymmetric Algorithms to Support Symmetric Cryptography

Source: ISACA, *CRISC Review Manual 6th Edition*, USA, 2015

---

[49] ISACA, *CRISC Review Manual 6th Edition*, USA, 2015

ELLIPTICAL CURVE CRYPTOGRAPHY

Although public key cryptography ensures message security, the long keys and mathematical problems it uses tend to be inefficient. A variant and more efficient form of public key cryptography is elliptical curve cryptography (ECC), which is gaining prominence as a method for increasing security while using minimum resources. It is believed that ECC demands less computational power and therefore offers more security per bit. For example, an ECC with a single bit key offers the same security as an RSA-based system with a 1,024-bit key.

ECC works well on networked computers requiring strong cryptography. However, it has some limitations such as bandwidth and processing power.

QUANTUM CRYPTOGRAPHY

Quantum cryptography is the next generation of cryptography that may solve some of the existing problems associated with current cryptographic systems, specifically the random generation and secure distribution of symmetric cryptographic keys. It is based on a practical application of the characteristics of the smallest "grains" of light (photons) and the physical laws governing their generation, propagation and detection.

DIGITAL SIGNATURE

A **digital signature** is an electronic identification of a person or entity created by using a public key algorithm. It serves as a way for the recipient to verify the integrity of the data and the identity of the sender. To verify the integrity of the data, a cryptographic hashing algorithm, called a checksum, is computed against the entire message or electronic document, which generates a small fixed string message, usually about 128 bits in length. This process, also referred to as a digital signature algorithm, creates a message digest (i.e., smaller extrapolated version of the original message).

Common types of message digest algorithms are SHA-256 and SHA-512. These algorithms are one-way functions, unlike private and public key encryption algorithms. The process of creating message digests cannot be reversed. They are meant for digital signature applications where a large electronic document or string of characters, such as word processor text, a spreadsheet, a database record, the content of a hard disk or a JPEG image has to be compressed in a secure manner before being signed with the private key. All digest algorithms take a message of arbitrary length and produce a 128-bit message digest.

The next step, which verifies the identity of the sender, is to encrypt the message digest using the sender's private key, which "signs" the document with the sender's digital signature for message authenticity. To decipher, the receiver would use the sender's public key, proving that the message could only have come from the sender. This process of sender authentication is known as nonrepudiation because the sender cannot later claim that they did not generate the message.

Once decrypted, the receiver will recompute the hash using the same hashing algorithm on the electronic document and compare the results with what was sent, to ensure the integrity of the message. Therefore, digital signature is a cryptographic method that ensures:
- **Data integrity**—Any change to the plaintext message would result in the recipient failing to compute the same message hash.
- **Authentication**—The recipient can ensure that the message has been sent by the claimed sender since only the claimed sender has the secret key.
- **Nonrepudiation**—The claimed sender cannot later deny generating and sending the message.

Digital signatures and public key encryption are vulnerable to man-in-the-middle attacks where the sender's digital signature private key and public key may be faked. To protect against such attacks, an independent sender authentication authority has been designed. The PKI performs the function of independently authenticating the validity of senders' digital signatures and public keys.

APPLICATIONS OF CRYPTOGRAPHIC SYSTEMS

The use of cryptosystems by applications, for example in email and Internet transactions, generally involves a combination of private/public key pairs, secret keys, hash functions and digital certificates. The purpose of applying these combinations is to achieve confidentiality, message integrity, or nonrepudiation by either the sender or recipient. The process generally involves the sender hashing the message into a message digest or pre-hash code for message integrity, which is encrypted using the sender's private key for authenticity, integrity and nonrepudiation (i.e., digital signature).

Using his/her secret key, the sender then will encrypt the message. Afterward, the secret key is encrypted with the recipient's public key, which has been validated through the recipient's digital certificate and provides message confidentiality. The process on the receiving end reverses what has been done by the sender. The recipient uses his/her private key to decrypt the sender's secret key. He/she uses this secret key to decrypt the message, to expose it. If the pre-hash code has been encrypted with the sender's private key, the recipient verifies its authenticity using the public key contained in the sender's digital certificate and decrypts the pre-hash code, which provides the nonrepudiation to the recipient of the sender's message. For integrity purposes, the recipient calculates a post-hash code, which should equal the pre-hash code. Specific examples of this method or related variants are described below.

Transport Layer Security (TLS)[50]—TLS is a cryptographic protocol that provides secure communications on the Internet. TLS is a session- or connection-layered protocol widely used for communication between browsers and web servers. Besides communication privacy, it also provides endpoint authentication. The protocols allow client-server applications to communicate in a way designed to prevent eavesdropping, tampering and message forgery.

TLS involves a number of basic phases:
- Peer negotiation for algorithm support
- Public-key, encryption-based key exchange and certificate-based authentication
- Symmetric cipher-based traffic encryption

During the first phase, the client and server negotiate which cryptographic algorithms will be used. Current implementations support the following choices:
- For public-key cryptography: RSA, Diffie-Hellman, DSA or Fortezza
- For symmetric ciphers: RC4, IDEA, Triple DES or AES
- For one-way hash functions: SHA-1 or SHA-2 (SHA-256)

TLS runs on layers above the TCP transport protocol and provides security to application protocols, even if it is most commonly used with HTTP to form Secure Hypertext Transfer Protocol (HTTPS). HTTPS is similar to HTTP, just with an encrypted session via TLS (or SSL) protocols. HTTPS serves to secure World Wide Web pages for applications. More, in electronic commerce, authentication may be used both in business-to-business (B-to-B) activities (for which both the client and the server are authenticated) and business-to-consumer (B-to-C) interaction (in which only the server is authenticated).

Besides TLS, SSL protocol is also widely used in real-world applications, even though its use is now deprecated as a significant vulnerability was discovered in 2014. TLS is a further development of SSL, but TLS and SSL are not interchangeable. Interoperability between SSL and TLS is impossible.

Secure Hypertext Transfer Protocol (HTTPS)—As an application layer protocol, HTTPS transmits individual messages or pages securely between a web client and server by establishing a TLS-type connection. Using the https:// designation in the URL instead of the standard http://, HTTPS directs the message to a secure port number rather than the default web port address. This protocol uses SSL secure features but does so as a message rather than as a session-oriented protocol.

---

[50] ISACA, *CISA Review Manual 26th Edition*, USA, 2015

Virtual Private Network (VPN)—A VPN is a secure private network that uses the public telecommunications infrastructure to transmit data. In contrast to a much more expensive system of owned or leased lines that can only be used by one company, VPNs are used by enterprises for both extranets and wide areas of intranets. Using encryption and authentication, a VPN encrypts all data that pass between two Internet points, maintaining privacy and security. Encryption is needed to make the connection virtually private. A popular VPN technology is IPSec, which commonly uses the DES, Triple DES or AES encryption algorithms.

IPSec—IPSec is used for communication among two or more hosts, two or more subnets, or hosts and subnets. This IP network layer packet security protocol establishes VPNs via transport and tunnel mode encryption methods. For the transport method, the data portion of each packet—referred to as the encapsulation security payload (ESP)—is encrypted to achieve confidentiality. In the tunnel mode, the ESP payload and its header are encrypted. To achieve nonrepudiation, an additional authentication header (AH) is applied.

In establishing IPSec sessions in either mode, security associations (SAs) are established. SAs define which security parameters should be applied between the communicating parties as encryption algorithms, keys, initialization vectors, life span of keys, etc.

To increase the security of IPSec, use asymmetric encryption via Internet Security Association and Key Management Protocol/Oakley (ISAKMP/Oakley), which allows the key management, use of public keys, negotiation, establishment, modification and deletion of SAs and attributes. For authentication, the sender uses digital certificates. The connection is made secure by supporting the generation, authentication and distribution of the SAs and cryptographic keys.

SSH—SSH is a client-server program that opens a secure, encrypted command-line shell session from the Internet for remote logon. Similar to a VPN, SSH uses strong cryptography to protect data, including passwords, binary files and administrative commands, transmitted between systems on a network. SSH is typically implemented by validating both parties' credentials via digital certificates. SSH is useful in securing Telnet and FTP services. It is implemented at the application layer, as opposed to operating at the network layer (IPSec implementation).

Secure Multipurpose Internet Mail Extensions (S/MIME)—S/MIME is a standard secure email protocol that authenticates the identity of the sender and receiver, verifies message integrity, and ensures the privacy of a message's contents, including attachments.

Secure Electronic Transactions (SET)—SET is a protocol developed jointly by VISA and MasterCard to secure payment transactions among all parties involved in credit card transactions. As an open system specification, SET is an application-oriented protocol that uses trusted third parties' encryption and digital signature processes, via a PKI of trusted third-party institutions, to address confidentiality of information, integrity of data, cardholder authentication, merchant authentication and interoperability.

PUBLIC KEY INFRASTRUCTURE

If an individual wants to send messages or electronic documents and sign them with a digital signature using a public key cryptographic system, how does the individual distribute the public key in a secure way? If the public key is distributed electronically, it could be intercepted and changed. To prevent this from occurring, a framework known as a PKI is used. PKI allows a trusted party to issue, maintain and revoke public key certificates.

PKI allows users to interact with other users and applications to obtain and verify identities and keys from trusted sources. The actual implementation of PKI varies according to specific requirements. Key elements of the infrastructure are as follows:
- **Digital certificates**—A digital certificate is composed of a public key and identifying information about the owner of the public key. The purpose of digital certificates is to associate a public key with the individual's identity in order to prove the sender's authenticity. These certificates are electronic documents, digitally signed by some trusted entity with its private key (transparent to users) that contains information about the individual and his or her public key. The process requires the sender to "sign" a document by attaching a digital certificate issued by a trusted entity. The receiver of the message and accompanying digital certificate relies on the public key of the

trusted third-party certificate authority (CA) (that is included with the digital certificate or obtained separately) to authenticate the message. The receiver can link the message to a person, not simply to a public key, because of their trust in this third party. The status and values of a current user's certificate should include:
– A distinguishing username
– An actual public key
– The algorithm used to compute the digital signature inside the certificate
– A certificate validity period

- **Certificate authority**—A CA is an authority in a network that issues and manages security credentials and public keys for message signature verification or encryption. The CA attests to the authenticity of the owner of a public key. The process involves a CA who makes a decision to issue a certificate based on evidence or knowledge obtained in verifying the identity of the recipient. As part of a PKI, a CA checks with a registration authority (RA) to verify information provided by the requestor of a digital certificate. If the RA verifies the requestor's information, the CA can then issue a certificate. Upon verifying the identity of the recipient, the CA signs the certificate with its private key for distribution to the user. Upon receipt, the user will verify the certificate signature with the CA's public key (e.g., commercial CAs such as VeriSign™ issue certificates through web browsers). The ideal CA is authoritative (someone that the user trusts) for the name or key space it represents. A certificate always includes the owner's public key, expiration date and the owner's information. Types of CAs may include:
  – Organizationally empowered, which have authoritative control over those individuals in their name space
  – Liability empowered, for example, choosing commercially available options (such as VeriSign) in obtaining a digital certificate. The CA is responsible for managing the certificate throughout its life cycle. Key elements or subcomponents of the CA structure include the certification practice statement (CPS), RAs and certificate revocation lists (CRLs).

- **Registration authority**—An RA is an authority in a network that verifies user requests for a digital certificate and tells the CA to issue it. An optional entity separate from a CA, an RA would be used by a CA with a very large customer base. CAs use RAs to delegate some of the administrative functions associated with recording or verifying some or all of the information needed by a CA to issue certificates or CRLs and to perform other certificate management functions. However, with this arrangement, the CA still retains sole responsibility for signing either digital certificates or CRLs. RAs are part of a PKI. The digital certificate contains a public key that is used to encrypt messages and verify digital signatures. If an RA is not present in the PKI structure established, the CA is assumed to have the same set of capabilities as those defined for an RA. The administrative functions that a particular RA implements will vary based on the needs of the CA, but must support the principle of establishing or verifying the identity of the subscriber. These functions may include the following:
  – Verifying information supplied by the subject (personal authentication functions)
  – Verifying the right of the subject to requested certificate attributes
  – Verifying that the subject actually possesses the private key being registered and that it matches the public key requested for a certificate (generally referred to as proof of possession [POP])
  – Reporting key compromise or termination cases where revocation is required
  – Assigning names for identification purposes
  – Generating shared secrets for use during the initialization and certificate pick-up phases of registration
  – Initiating the registration process with the CA on behalf of the subject end entity
  – Initiating the key recovery processing
  – Distributing the physical tokens (such as smart cards) containing the private keys
  – Certificate revocation list—The CRL is an instrument for checking the continued validity of the certificates for which the CA has responsibility. The CRL details digital certificates that are no longer valid because they were revoked by the CA. The time gap between two updates is critical and is also a risk in digital certificates verification.
  – Certification practice statement—CPS is a detailed set of rules governing the CA's operations. It provides an understanding of the value and trustworthiness of certificates issued by a given CA in terms of the following:
    - The controls that an organization observes
    - The method it uses to validate the authenticity of certificate applicants
    - The CA's expectations of how its certificates may be used

STORED DATA

Encryption is an effective and increasingly practical way to restrict access to confidential information while in storage. The traditional protection method—a password—has inherent weaknesses and, in many cases, is easily guessable. Access control lists (ACLs) that define who has access are also effective, but they often have to be used in conjunction with operating systems or applications. Further, ACLs cannot prevent improper use of information by systems administrators, as the latter can have total control of a computer. Encryption can fill the security gap, and it can also protect data from hackers who, by means of malicious software, can obtain systems administration rights. Encryption also helps to protect data when a computer or a disk falls into the wrong hands. Many email encryption programs can also be applied to stored data. There are also some encryption products that focus on file protection for computers and mobile smart devices.

ENCRYPTION RISK AND KEY PROTECTION

The security of encryption methods relies mainly on the secrecy of keys. In general, the more a key is used, the more vulnerable it will be to compromise. For example, password cracking tools for today's microcomputers can brute force every possible key combination for a cryptographic hashing algorithm with a 40-bit key in a matter of a few hours.

The randomness of key generation is also a significant factor in the ability to compromise a key. When passwords are tied into key generation, the strength of the encryption algorithm is diminished, particularly when common words are used. This significantly reduces the key space combinations to search for the key. For example, an eight-character password is comparable to a 32-bit key. When encrypting keys based on passwords, a password that lacks randomness will diminish a 128-bit encryption algorithm's capabilities. Therefore, it is essential that effective password syntax rules are applied and easily guessed passwords are prohibited.

SECTION 3—KNOWLEDGE CHECK

1. Select all that apply. The Internet perimeter should:
   A. detect and block traffic from infected internal end points.
   B. eliminate threats such as email spam, viruses and worms.
   C. format, encrypt and compress data.
   D. control user traffic bound toward the Internet.
   E. monitor internal and external network ports for rogue activity.

2. The _____ layer of the OSI model ensures that data are transferred reliably in the correct sequence, and the _____ layer coordinates and manages user connections.
   A. Presentation, data link
   B. Transport, session
   C. Physical, application
   D. Data link, network

3. Choose three. The key benefits of the DMZ system are:
   A. DMZs are based on logical rather than physical connections.
   B. an intruder must penetrate three separate devices.
   C. private network addresses are not disclosed to the Internet.
   D. excellent performance and scalability as Internet usage grows.
   E. internal systems do not have direct access to the Internet.

4. Which of the following best states the role of encryption within an overall cybersecurity program?
   A. Encryption is the primary means of securing digital assets.
   B. Encryption depends upon shared secrets and is therefore an unreliable means of control.
   C. A program's encryption elements should be handled by a third-party cryptologist.
   D. Encryption is an essential but incomplete form of access control.

5. The number and types of layers needed for defense in depth are a function of:
   A. asset value, criticality, reliability of each control and degree of exposure.
   B. threat agents, governance, compliance and mobile device policy.
   C. network configuration, navigation controls, user interface and VPN traffic.
   D. isolation, segmentation, internal controls and external controls.

See answers in Appendix C.

Page intentionally left blank

Section 4:
Security of Networks, Systems, Applications and Data

Topics covered in this section include:
1. Process controls—risk assessments
2. Process controls—vulnerability management
3. Process controls—penetration testing
4. Network security
5. Operating system security
6. Application security
7. Data security

Page intentionally left blank

TOPIC 1—PROCESS CONTROLS—RISK ASSESSMENTS

As previously mentioned, **risk** is defined as the possibility of loss of a digital asset resulting from a threat exploiting a **vulnerability**. Each of these attributes of risk must be analyzed to determine an organization's particular risk. The process of doing this analysis is called a **cyberrisk assessment.**

While every risk assessment methodology has different nuances and approaches, most have three common inputs: asset identification, threat assessment and vulnerability assessment, as shown in figure 4.1.

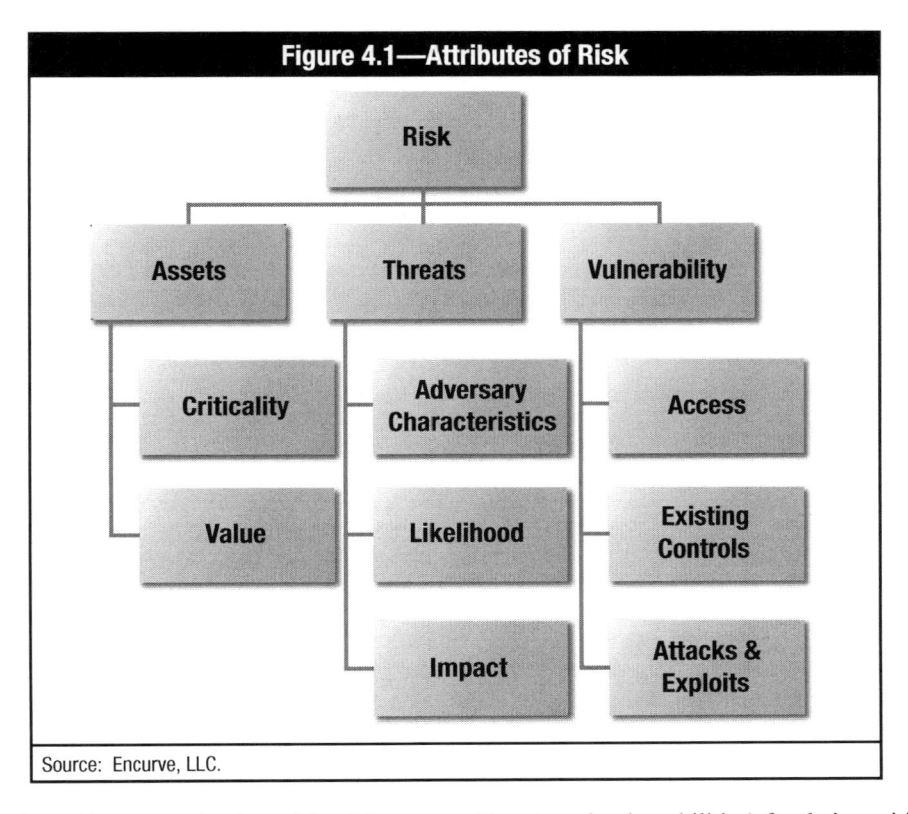

Figure 4.1—Attributes of Risk

- Risk
  - Assets
    - Criticality
    - Value
  - Threats
    - Adversary Characteristics
    - Likelihood
    - Impact
  - Vulnerability
    - Access
    - Existing Controls
    - Attacks & Exploits

Source: Encurve, LLC.

This process begins with an examination of the risk sources (threats and vulnerabilities) for their positive and negative consequences.

After evaluating each of these attributes, risk can be ranked according to likelihood and impact. Information used to estimate impact and likelihood usually comes from:
- Past experience or data and records (e.g., incident reporting)
- Reliable practices, international standards or guidelines
- Market research and analysis
- Experiments and prototypes
- Economic, engineering or other models
- Specialist and expert advice

Section 4: Security of Networks, Systems, Applications and Data

Finally, existing controls and other mitigation strategies are evaluated to determine the level and effectiveness of risk mitigation currently in place and identify deficiencies and gaps that require attention. A risk response workflow is shown in **figure 4.2**.

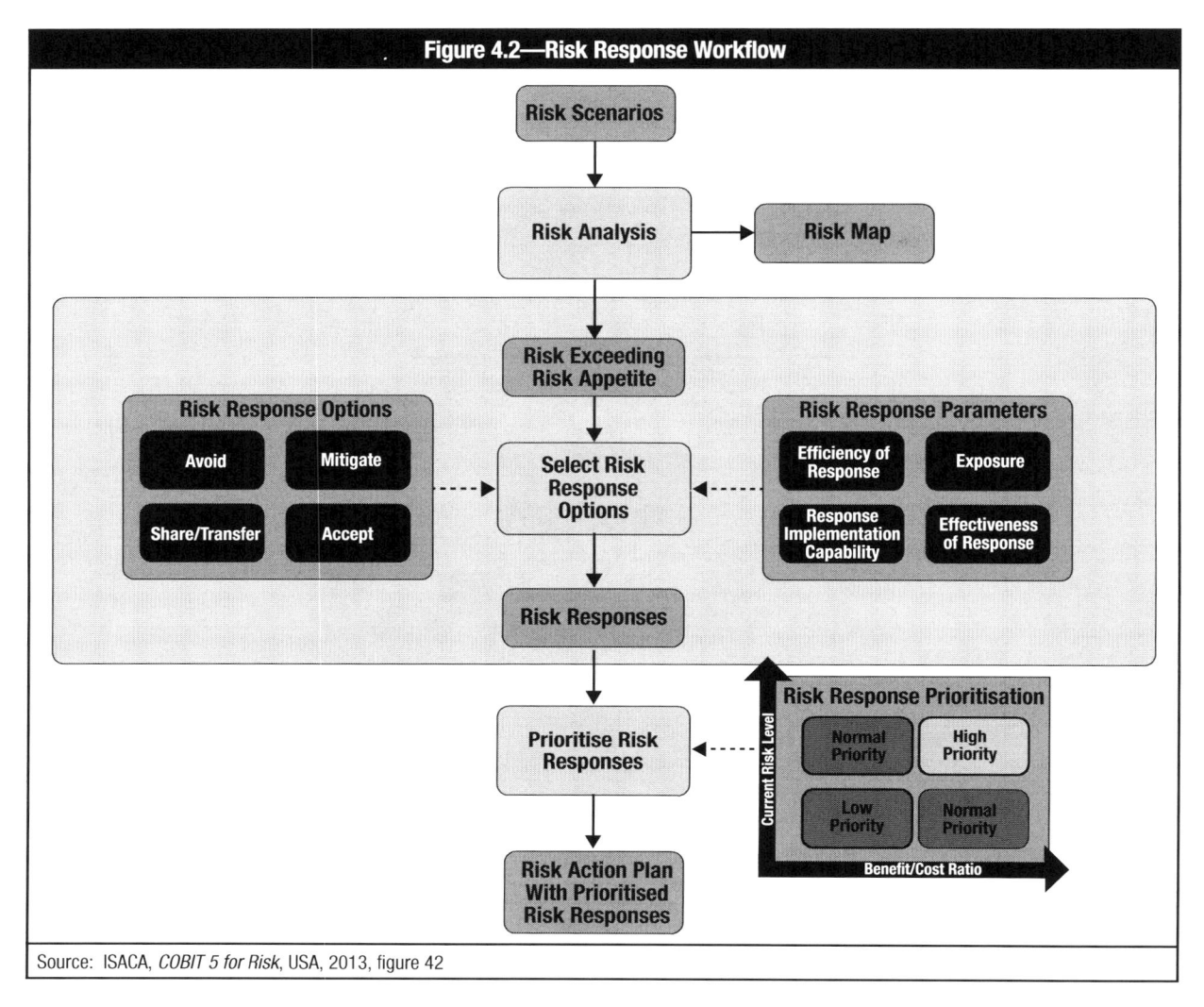

Figure 4.2—Risk Response Workflow

Source: ISACA, *COBIT 5 for Risk*, USA, 2013, figure 42

It is critical for every cybersecurity professional to understand the basic concepts and nomenclature of the risk assessment process. If risk is not properly analyzed, the implementation of security is left to guesswork. In the following sections, common risk assessment processes will be covered in more detail.

RISK ANALYSIS

As stated previously, there are many methods used to bring the data collected on assets, threats and vulnerabilities together and analyze them to determine risk. Most rely on some process to pair and prioritize likelihoods and impacts. Additionally, risk analyses can be oriented toward one of the inputs, making the risk assessment asset-oriented, threat-oriented or vulnerability-oriented, as shown in **figure 4.3**.[51]

Orientation	Description
Asset	Important assets are defined first, and then potential threats to those assets are analyzed. Vulnerabilities that may be exploited to access the asset are identified.
Threat	Potential threats are determined first, and then threat scenarios are developed. Based on the scenarios, vulnerabilities and assets of interest to the adversary are determined in relation to the threat.
Vulnerability	Vulnerabilities and deficiencies are identified first, then the exposed assets, and then the threat events that could be taken advantage of are determined.

Figure 4.3—Risk Assessment Orientations

No one analysis orientation is better than the other; however, each has a bias that, if not considered, could weaken the analysis process resulting in some risk not being identified or properly prioritized. Some organizations will perform risk assessments from more than one orientation to compensate for the potential bias and generate a more thorough analysis.

EVALUATING SECURITY CONTROLS

After risk is identified and prioritized, existing controls should be analyzed to determine their effectiveness in mitigating the risk. This analysis will result in a final risk ranking based on risk that has adequate controls, inadequate controls and no controls.

A very important criterion in control selection and evaluation is that the cost of the control (including its operation) should not exceed value of the asset it is protecting.

RISK ASSESSMENT SUCCESS CRITERIA

Choosing the exact method of analysis, including qualitative or quantitative approaches and determining the analysis orientation, takes considerable planning and knowledge of specific risk assessment methodologies. To be successful, the risk assessment process should fit the goals of the organization, adequately address the environment being assessed and use assessment methodologies that fit the data that can be collected.

The scope of the assessment must be clearly defined and understood by everyone involved in the risk assessment process. The process should be simple enough to be completed within the scope and time frame of the project yet rigorous enough to produce meaningful results.

It is important to understand the organization's unique risk appetite and cultural considerations when performing a risk assessment. Cultural aspects can have a significant impact on risk management. For example, financial institutions have more formal, regulated cultures where selection and implementation of stringent controls is acceptable, whereas a small entrepreneurial start-up may see some types of security controls as a hindrance to business.

Finally, risk assessment is not a one-off process. No organization is static; technology, business, regulatory and statutory requirements, people, vulnerabilities and threats are continuously evolving and changing. Therefore, successful risk assessment is an ongoing process to identify new risk and changes to the characteristics of existing and known risk.

---

[51] National Institute of Standards and Technology (NIST), *Special Publication 800-30, Guide for Conducting Risk Assessments*, USA, September 2012

MANAGING RISK

For risk that has inadequate or no controls, there are many options to address each risk, as shown in **figure 4.4**.

Figure 4.4—Risk Response Strategy	
Risk Response	**Description**
Risk Reduction	The implementation of controls or countermeasures to reduce the likelihood or impact of a risk to a level within the organization's risk tolerance.
Risk Avoidance	Risk can be avoided by not participating in an activity or business.
Risk Transfer or Sharing	Risk can be transferred to a third party (e.g., insurance) or shared with a third party via contractual agreement.
Risk Acceptance	If the risk is within the organization's risk tolerance or if the cost of otherwise mitigating the risk is higher than the potential loss, then an organization can assume the risk and absorb any losses.

What strategy an organization chooses depends on many different things such as regulatory requirements, culture, mission, ability to mitigate risk and risk tolerance.

USING THE RESULTS OF THE RISK ASSESSMENT

Risk assessment results are used for a variety of security management functions. These results need to be evaluated in terms of the organization's mission, risk tolerance, budgets and other resources, and cost of mitigation. Based on this evaluation, a mitigation strategy can be chosen for each risk and appropriate controls and countermeasures can be designed and implemented.

Risk assessment results can also be used to communicate the risk decisions and expectations of management throughout the organization through policies and procedures.

Finally, risk assessments can be used to identify areas where incident response capabilities need to be developed to quickly detect and respond to inherent or residual risk or where security controls cannot adequately address the threat.

TOPIC 2—PROCESS CONTROLS—VULNERABILITY MANAGEMENT

Vulnerabilities are continuously being discovered and organizations must be constantly vigilant in identifying them and quickly remediating.

Organizations need to identify and assess vulnerabilities to determine the threat and potential impact and to determine the best course of action in addressing each vulnerability. Vulnerabilities can be identified by information provided by software vendors (e.g., through the release of patches and updates) and by utilizing processes and tools that identify known vulnerabilities in the organization's specific environment. The two most common techniques are vulnerability scanning and penetration testing.

VULNERABILITY MANAGEMENT

Vulnerability management starts by understanding the IT assets and where they reside—both physically and logically. This can be done by maintaining an asset inventory that details important information about each asset such as location (physical or logical), criticality of the asset, the organizational owner of the asset and the type of information the asset stores or processes.

VULNERABILITY SCANS

Vulnerability scanning is the process of using proprietary or open source tools to search for known vulnerabilities. Often the same tools used by adversaries to identify vulnerabilities are used by organizations to locate vulnerabilities proactively.

There are many forms of vulnerability assessment tools. Several Linux distributions (e.g., Kali Linux) supply open source tools. Commercial tools (e.g., Core Impact, Nessus®, Nexpose®) are often used to scan IT infrastructure, web application, databases or a mix of them. The licensed updates of the vulnerability rule base fall into two categories: host-based and network-based. Naming every tool is impractical because individual needs and budgets vary. Likewise, higher cost does not always equate to greater functionality, and tools can be found that are either free or free to try. Tools should be researched and selected based on corporate needs and return on investment, keeping in mind that combinations of tools often provide greater insight to your network's security posture.

Vulnerability scans should be conducted regularly to identify new vulnerabilities and ensure that previously identified vulnerabilities have been properly corrected.

Section 4: Security of Networks, Systems, Applications and Data

VULNERABILITY ASSESSMENT

The simplest definition of a vulnerability is an exploitable weakness that results in a loss. The method used to take advantage of a vulnerability is called an exploit. Vulnerabilities can occur in many different forms and at different architectural levels (e.g., physical, operating system, application). **Figure 4.5** provides a list of common types of vulnerabilities.

Figure 4.5—Common Types of Vulnerabilities		
Type of Vulnerability	**Cause**	**Cybersecurity Examples**
Technical	Errors in design, implementation, placement or configuration	• Coding errors • Inadequate passwords • Open network ports • Lack of monitoring
Process	Errors in operation	• Failure to monitor logs • Failure to patch software
Organizational	Errors in management, decision, planning or from ignorance	• Lack of policies • Lack of awareness • Failure to implement controls
Emergent	Interactions between, or changes in, environments	• Cross-organizational failures • Interoperability errors • Implementing new technology

It is important to analyze vulnerabilities in the context of how they are exploited, and both vulnerabilities and exploits need to be considered in vulnerability assessments. Vulnerabilities and exploits can be identified in many ways. At a technical level, automated tools (both proprietary and open source) can be used to identify common vulnerabilities in computer and network implementations and configurations. Other vulnerability analysis tools include open source and proprietary sources such as SANS, MITRE and OWASP, software vendors, historical incidents, etc.

REMEDIATION

After vulnerabilities are identified and assessed, appropriate remediation can take place to mitigate or eliminate the vulnerability. Most often, remediation will be through a patch management process but may also require reconfiguration of existing controls or addition of new controls.

REPORTING AND METRICS

Vulnerability management includes tracking vulnerabilities and the remediation efforts to mitigate them. This provides a clear opportunity to provide good qualitative metrics to the organization's management on the numbers and types of vulnerabilities, the potential impacts and the effort needed to mitigate them.

TOPIC 3—PROCESS CONTROLS—PENETRATION TESTING

Penetration testing includes identifying existing vulnerabilities and then using known exploit methods to:
• Confirm exposures.
• Assess the level of effectiveness and quality of existing security controls.
• Identify how specific vulnerabilities expose IT resources and assets.
• Ensure compliance.

Because penetration testing simulates actual attacks, they must be planned carefully. Failure to do so may result in ineffective results, negative impact on or damage to the organization's IT infrastructure, potential liability, or criminal prosecution. Several considerations are important prior to any penetration testing:
• Clearly define the scope of the test including what systems or networks are in and out of scope, the type of exploits that may be used, and the level of access allowed. These exploits can include network, social engineering, web, mobile application and other kinds of testing.
• Gather explicit, written permission from the organization authorizing the testing. This is the only accepted industry standard that distinguishes the service as authorized and legal.
• Ensure testers implement "Do no harm" procedures to ensure no assets are harmed, such as deletions, denial-of-service (DoS) or other negative impacts.
• Put in place communication and escalation plans for the organization and testers to communicate quickly during the tests.

PENETRATION TESTERS

Penetration testing requires specialized knowledge of vulnerabilities, exploits, IT technology and the use of testing tools. It should not be performed by untrained or unqualified practitioners. Any penetration tests should be carefully planned to mitigate the risk of causing a service outage, and the results require careful interpretation and elimination of false positives.

One commonly held belief is that penetration tests are not high priority because they are typically successful in finding a hole in an organization's defense. The thought is that if an organization can design a successful penetration test (i.e., create/identify the vulnerability), than money is better spent creating the fix rather than testing the vulnerability.

Penetration testing can be covert (the general IT staff do not know the testing is going to take place) so that the reactions of the organization to detect and respond are also tested. Also, penetration testing can be external, from outside the organization, or internal, starting from a system behind the organization's firewall.[52]

PENETRATION TESTING FRAMEWORKS

Penetration testing should use a framework to deliver repeatability, consistency and high quality in various kinds of security tests.

Penetration testing frameworks include:[53]
• **PCI Penetration Testing Guide**—Provides a good introduction to testing tools
• **Penetration Testing Execution Standard**—Provides hands-on technical guidance on penetration testing
• **Penetration Testing Framework**—Provides a comprehensive guide to penetration testing and testing tools
• **Information Systems Security Assessment Framework (ISSAF)**—Provides comprehensive penetration technical guidance
• **Open Source Security Testing Methodology Manual (OSSTMM)**—Provides a methodology for testing operational security and can support ISO 27001

---

[52] Encurve, LLC.
[53] Open Web Application Security Project (OWASP), *Penetration testing methodologies*, www.owasp.org/index.php/Penetration_testing_methodologies

PENETRATION TESTING COMMON PHASES

Penetration testing can be divided into four common main phases, as shown in **figure 4.6**.

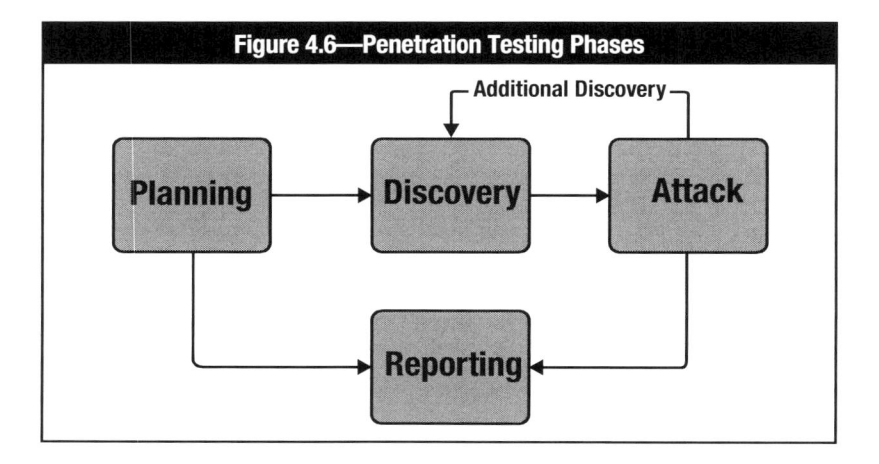

Figure 4.6—Penetration Testing Phases

Planning → Discovery → Attack
Additional Discovery (loop between Discovery and Attack)
Planning, Discovery, and Attack all feed into Reporting.

The phases include:
1. **Planning**—In the planning phase, the goals are set, the scope is defined, and the test is approved and documented by management. The scope determines if the penetration test is internal or external, limited to certain types of attacks or limited to certain networks or assets.
2. **Discovery**—In the discovery phase, the penetration tester gathers information by conducting research on the organization and scans the networks for port and service identification. Techniques used to gather information include:
   – DNS interrogation, WHOIS queries and network sniffing to discover host name and IP address information
   – Search web servers and directory servers for employee names and contact information
   – Banner grabbing for application and service information
   – NetBIOS enumeration for system information
   – Dumpster diving and physical walk-throughs of the facilities to gather additional information
   – Use of online Internet infrastructure search tools, such as Shodan[54], to passively profile exposed systems and services
   – Social engineering, such as posing as a help desk agent and asking for passwords, posing as a user and calling the help desk to reset passwords, or sending phishing emails

   A vulnerability assessment is also conducted during the discovery phase. This involves comparing the services, applications and operating systems of the scanned host against vulnerability databases.
3. **Attack**—The attack phase is the process of verifying previously identified vulnerabilities by attempting to exploit them. Metasploit® hosts a public database of quality-assured exploits. They rank exploits for safe testing.

Sometimes exploit attempts do not provide the tester with access, but they do give the tester additional information about the target and its potential vulnerabilities. If a tester is able to exploit a vulnerability, they can install more tools on the system or network to gain access to additional systems or resources.

A payload is the piece of software that lets a user control a computer system after it has been exploited. The payload is typically attached to and delivered by the exploit. Metasploit's most popular payload is called Meterpreter, which enables a user to upload and download files from the system, take screenshots and collect password hashes. The discovery and attack phases are illustrated in **figure 4.7**.

---
[54] More information is available at *www.shodan.io/*.

4. **Reporting:** The reporting phase occurs simultaneously with the other phases. An assessment plan is developed during the planning phase. Logs are kept during the discovery and attack phases. And, at the conclusion of the penetration test, a report is developed to describe the vulnerabilities identified, assign risk ratings and provide mitigation plans.

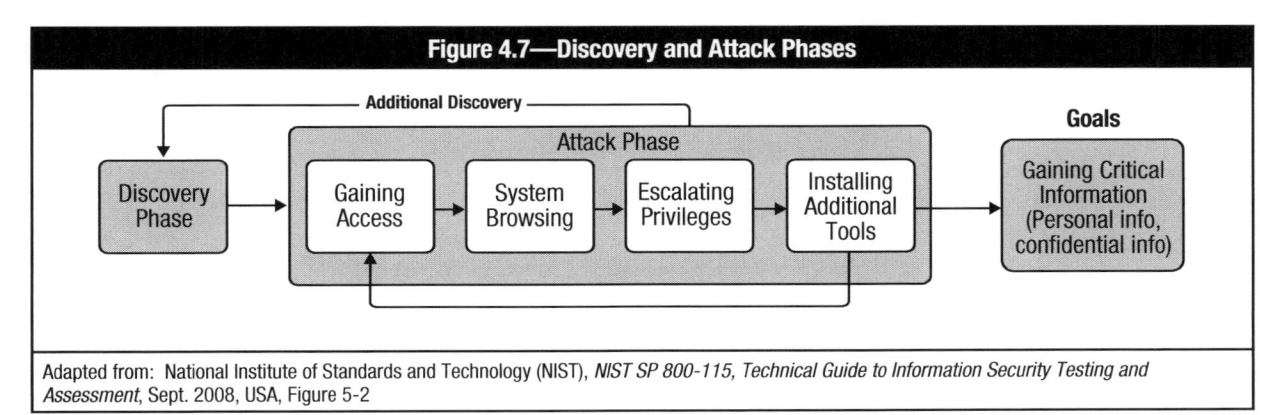

Figure 4.7—Discovery and Attack Phases

Discovery Phase → [Attack Phase: Gaining Access → System Browsing → Escalating Privileges → Installing Additional Tools] → Goals: Gaining Critical Information (Personal info, confidential info)

Additional Discovery loops back from the Attack Phase to the Discovery Phase.

Adapted from: National Institute of Standards and Technology (NIST), *NIST SP 800-115, Technical Guide to Information Security Testing and Assessment*, Sept. 2008, USA, Figure 5-2

Page intentionally left blank

TOPIC 4—NETWORK SECURITY

NETWORK MANAGEMENT

Network management is the process of assessing, monitoring and maintaining network devices and connections. The recommended functions of network management are listed within a model made up of five functional areas (FCAPS). These five functional areas, introduced by the International Organization for Standardization (ISO) and further developed by the International Telecommunication Union (ITU), are shown in **figure 4.8** and as follows:

- **Fault management**—Detect, isolate, notify and correct faults encountered in the network. This category analyzes traffic, trends, SNMP polls and alarms for automatic fault detection.
- **Configuration management**—Configuration aspects of network devices include configuration file management, inventory management and software management.
- **Accounting management**—Usage information of network resources.
- **Performance management**—Monitor and measure various aspects of performance metrics so that acceptable performance can be maintained. This includes response time, link utilization and error rates. Administrators can monitor trends and set threshold alarms.
- **Security management**—Provide access to network devices and corporate resources to authorized individuals. This category focuses on authentication, authorization, firewalls, network segmentation, IDS and notifications of attempted breaches.

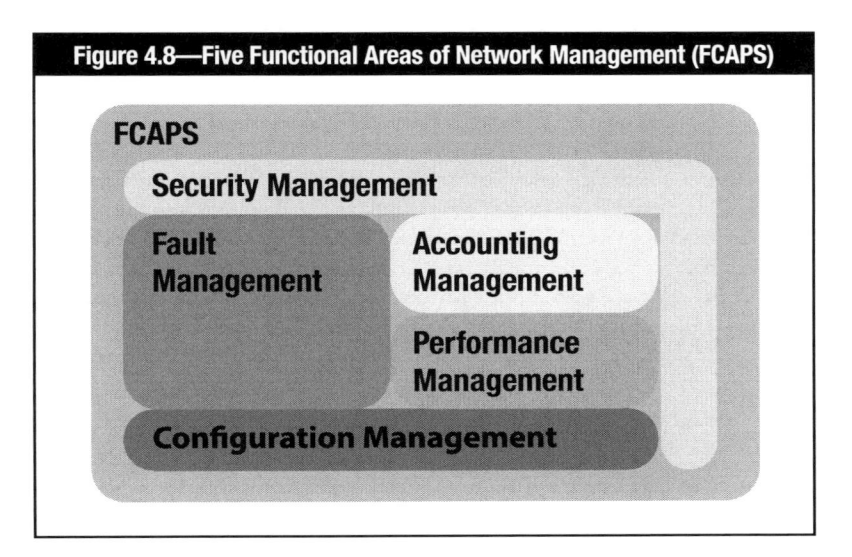

Figure 4.8—Five Functional Areas of Network Management (FCAPS)

FCAPS
- Security Management
- Fault Management
- Accounting Management
- Performance Management
- Configuration Management

LOCAL AREA NETWORK (LAN)[55]

A LAN covers a small, local area—from a few devices in a single room to a network across a few buildings. The increase in reasonably priced bandwidth has reduced the design effort required to provide cost-effective LAN solutions for organizations of any size.

New LANs are almost always implemented using switched Ethernet (802.3). Twisted-pair cabling (100-Base-T or better and wireless LANs [WLANs]) connects floor switches to the workstations and printers in the immediate area. Floor switches can be connected to each other with 1000-Base-T or fiber-optic cabling. In larger organizations, the floor switches may be connected to larger, faster switches whose purpose is to properly route the switch-to-switch data.

---

[55] ISACA, *CISA Review Manual 26th Edition*, USA, 2015

As LANs get larger and traffic increases, the requirement to carefully plan the logical configuration of the network becomes more and more important. Network planners need to be highly skilled and very knowledgeable. Their tools include traffic monitors that allow them to monitor traffic volumes on critical links. Tracking traffic volumes, error rates and response times is every bit as important on larger LANs as it is on distributed servers and mainframes.

LAN COMPONENTS[56]

Components commonly associated with LANs are repeaters, hubs, bridges, switches and routers:
- **Repeaters**—Physical layer devices that extend the range of a network or connect two separate network segments together. Repeaters receive signals from one network segment and amplify (regenerate) the signal to compensate for signals (analog or digital) that are distorted due to a reduction of signal strength during transmission (i.e., attenuation). Wi-Fi repeaters for home usage are currently popular.
- **Hubs**—Physical layer devices that serve as the center of a star-topology network or a network concentrator. Hubs can be active (if they repeat signals sent through them) or passive (if they merely split signals).
- **Layer 2 switches**—Layer 2 switches are data link level devices that can divide and interconnect network segments and help to reduce collision domains in Ethernet-based networks. Furthermore, switches store and forward frames, filtering and forwarding packets among network segments, based on Layer 2 MAC source and destination addresses, as bridges and hubs do at the data link layer. Switches, however, provide more robust functionality than bridges, through use of more sophisticated data link layer protocols, which are implemented via specialized hardware called application-specific integrated circuits (ASICs). The benefits of this technology are performance efficiencies gained through reduced costs, low latency or idle time, and a greater number of ports on a switch with dedicated high-speed bandwidth capabilities. Switches are also applicable in WAN technology specifications.
- **Routers**—Similar to bridges and switches in that they link two or more physically separate network segments. The network segments linked by a router, however, remain logically separate and can function as independent networks. Routers operate at the OSI network layer by examining network addresses (i.e., routing information encoded in an IP packet). By examining the IP address, the router can make intelligent decisions to direct the packet to its destination. Routers differ from switches operating at the data link layer in that they use logically based network addresses, use different network addresses/segments off all ports, block broadcast information, block traffic to unknown addresses, and filter traffic based on network or host information. Routers are often not as efficient as switches because they are generally software-based devices and they examine every packet coming through, which can create significant bottlenecks within a network. Therefore, careful consideration should be taken as to where routers are placed within a network. This should include leveraging switches in network design as well as applying load balancing principles with other routers for performance efficiency considerations.
- **Layer 3 and 4 switches**—Advances in switch technology have also provided switches with operating capabilities at Layer 3 and Layer 4 of the OSI reference model.
  – A Layer 3 switch goes beyond Layer 2, acting at the network layer of the OSI model like a router. The Layer 3 switch looks at the incoming packet's networking protocol (e.g., IP). The switch compares the destination IP address to the list of addresses in its tables, to actively calculate the best way to send a packet to its destination. This creates a "virtual circuit" (i.e., the switch has the ability to segment the LAN within itself and will create a pathway between the receiving and the transmitting device to send the data). It then forwards the packet to the recipient's address. This provides the added benefit of reducing the size of network broadcast domains. Broadcast domains should be limited or aligned with business functional areas/workgroups within an organization, to reduce the risk of information leakage to those without a need to know, where systems can be targeted and their vulnerabilities exploited. The major difference between a router and a Layer 3 switch is that a router performs packet switching using a microprocessor, whereas a Layer 3 switch performs the switching using application ASIC hardware.
  – A Layer 4 switch allows for policy-based switching. With this functionality, Layer 4 switches can off-load a server by balancing traffic across a cluster of servers, based on individual session information and status.
- **Layer 4-7 switches**—Also known as content-switches, content services switches, web-switches or application-switches. They are typically used for load balancing among groups of servers. Load balancing can be based on HTTP, HTTPS and/or VPN, or for any application TCP/IP traffic using a specific port. Content switches can also be used to perform standard operations such as SSL encryption/decryption to reduce the load on the servers receiving the traffic, and to centralize the management of digital certificates.

[56] *Ibid*

Section 4: Security of Networks, Systems, Applications and Data

LAN/WAN SECURITY

LANs and WANs are particularly susceptible to people and virus-related threats because of the large number of people who have access rights.

The administrative and control functions available with network software might be limited. Software vendors and network users have recognized the need to provide diagnostic capabilities to identify the cause of problems when the network goes down or functions in an unusual manner. The use of logon IDs and passwords with associated administration facilities is only becoming standard now (i.e., application aware networks). Read, write and execute permission capabilities for files and programs are options available with some network operating system versions, but detailed automated logs of activity (audit trails) are seldom found on LANs. Fortunately, newer versions of network software have significantly more control and administration capabilities.

LANs can represent a form of decentralized computing. Decentralized local processing provides the potential for a more responsive computing environment; however, organizations do not always give the opportunity to efficiently develop staff to address the technical, operational and control issues that the complex LAN technology represents. As a result, local LAN administrators frequently lack the experience, expertise and time to effectively manage the computing environment.

NETWORK ACCESS CONTROL (NAC)

Network access control (NAC) aims to control the access to a network using policies that describe how devices can secure access to network nodes when they first try to access a network. Some features include integrating an automatic remediation process that fixes noncompliant nodes before access is allowed and enabling network infrastructure to work with back office services and end-user computing to ensure that the network is secure prior to allowing access.

LAN RISK AND ISSUES

LANs facilitate the storage and retrieval of programs and data used by a group of people. LAN software and practices also need to provide for the security of these programs and data. Unfortunately, most LAN software provides a low level of security. The emphasis has been on providing capability and functionality rather than security. As a result, risk associated with use of LANs includes:
- Loss of data and program integrity through unauthorized changes
- Lack of current data protection through inability to maintain version control
- Exposure to external activity through limited user verification and potential public network access from dial-in connections
- Virus and worm infection
- Improper disclosure of data because of general access rather than need-to-know access provisions
- Violation of software licenses by using unlicensed or excessive numbers of software copies
- Illegal access by impersonating or masquerading as a legitimate LAN user
- Internal user's sniffing (obtaining seemingly unimportant information from the network that can be used to launch an attack such as network address information)
- Internal user's spoofing (reconfiguring a network address to pretend to be a different address)
- Destruction of the logging and auditing data

The LAN security provisions available depend on the software product, product version and implementation. Commonly available network security administrative capabilities include:
- Declaring ownership of programs, files and storage
- Limiting access to a read-only basis
- Implementing record and file locking to prevent simultaneous update
- Enforcing user ID/password sign-on procedures, including the rules relating to password length, format and change frequency
- Using switches to implement port security policies rather than hubs or unmanageable routers, which will prevent unauthorized hosts, with unknown MAC addresses, from connecting to the LAN
- Encrypting local traffic using IPSec (IP security) protocol

The use of these security procedures requires administrative time to implement and maintain. Network administration is often inadequate, providing global access because of the limited administrative support available when limited access is appropriate.

WIRELESS[57]

Wireless technologies, in the simplest sense, enable one or more devices to communicate without physical connections (i.e., without requiring network or peripheral cabling). Wireless is a technology that enables organizations to adopt ebusiness solutions with tremendous growth potential. Wireless technologies use radio frequency transmissions/electromagnetic signals through free space as the means for transmitting data, whereas wired technologies use electrical signals through cables. Wireless technologies range from complex systems (such as wireless wide area networks [WWANs], wireless local area networks [WLANs] and cell phones) to simple devices (such as wireless headphones, microphones, and other devices that do not process or store information). They also include Bluetooth® devices with a miniature radio frequency transceiver and infrared devices, such as remote controls, some cordless computer keyboards and mice, and wireless Hi-Fi stereo headsets, all of which require a direct line of sight between the transmitter and the receiver to close the link.

However, going wireless introduces new elements that must be addressed. For example, existing applications may need to be retrofit to make use of wireless interfaces. Also, decisions need to be made regarding general connectivity—to facilitate the development of completely wireless mobile applications or other applications that rely on synchronization of data transfer between mobile computing systems and corporate infrastructure. Other issues include narrow bandwidth, the lack of a mature standard, and unresolved security and privacy issues.

Wireless networks serve as the transport mechanism between devices, and among devices and the traditional wired networks. Wireless networks are many and diverse but are frequently categorized into four groups based on their coverage range:
- WANs
- LANs
- Wireless personal area networks (WPANs)
- Wireless ad hoc networks

WIRELESS LOCAL AREA NETWORKS (WLAN)[58]

WLANs allow greater flexibility and portability than traditional wired LANs. Unlike a traditional LAN, which requires a wire to connect a user's computer to the network, a WLAN connects computers, tablets, smartphones and other components to the network using an access point device. An access point, or wireless networking hub, communicates with devices equipped with wireless network adaptors within a specific range of the access point; it connects to a wired Ethernet LAN via an RJ-45 port. Access point devices typically have coverage areas of up to 300 feet (approximately 100 meters). This coverage area is called a cell or range. Users move freely within the cell with their laptop or other network devices. Access point cells can be linked together to allow users to roam within a building or between buildings. WLAN includes 802.11, HyperLAN, HomeRF and several others. WLANs are commonly referred to as Wi-Fi hotspots.

WLAN technologies conform to a variety of standards and offer varying levels of security features. The principal advantages of standards are to encourage mass production and to allow products from multiple vendors to interoperate. The most useful standard used currently is the IEEE 802.11 standard.

802.11 refers to a family of specifications for WLAN technology. 802.11 specifies an over-the-air interface between a wireless client and a base station or between two wireless clients.

---

[57] *Ibid*
[58] *Ibid*

WIRELESS NETWORK PROTECTIONS

Wireless data transmission is subject to a higher risk of interception than wired traffic, in the same way that it is easier to intercept calls made from mobile phones than calls from landline telephones. There is no need to manually tap into the connection, but rather remote tools can be used to intercept the connection covertly. Wireless transmission of confidential information should be protected with strong encryption. An insecure wireless connection exposes users to eavesdropping, which can lead to the exposure of confidential information, intercepted messages or abused connections. Examples include:

- Email can be intercepted and read or changed.
- Hackers can replace a user's credential with false information that leads to the destination server rejecting the user's access attempts, thereby causing DoS.
- An unauthorized person can log on to a wireless network that is not secure and use its resources, including free connectivity to the Internet.

IEEE 802.11's Wired Equivalent Privacy (WEP) encryption uses symmetric, private keys, which means the end user's radio-based network interface controller (NIC) and access point must have the same key. This leads to periodic difficulties distributing new keys to each NIC. As a result, keys remain unchanged on networks for extended times. With static keys, several hacking tools easily break through the relatively weak WEP encryption mechanisms.

Wireless security standards continue to evolve. The most commonly used method for wireless local area networks is 802.11i (WPA2) and Wi-Fi Protected Access (WPA), which uses dynamic keys and can use an authentication server with credentials to increase protection against hackers.

WEP and WPA comply with the evolving versions of the 802.11 wireless standard specified by IEEE, with WPA being compatible with more advanced versions of 802.11, even WPA has shortcomings. The key is protected with a passphrase that does not have a rigorously enforced length. WPA is a subset of the developing 802.11i standard. The full standard calls for enhanced security by implementing AES, which many vendors have introduced into their devices as an option.

WPA and WPA V2 (preferred) are applicable to most wireless networks and commonly used in networks that involve PCs. Messages transmitted using portable wireless devices should also be protected with encryption and, where possible, VPN methods can be used to provide additional security. For example, the BlackBerry® Enterprise Server model integrates the device with corporate email and uses Triple DES to encrypt information between the BlackBerry unit and a corporate mail server.

Public keys are also used in mobile devices. ECC is widely used on smart cards and is increasingly deployed for cell phones. ECC is suited for small devices because the algorithm, by combining plane geometry with algebra, can achieve stronger authentication with smaller keys compared to traditional methods, such as RSA, which primarily use algebraic factoring. Smaller keys are more suitable to mobile devices; however, some would argue that ECC is not as rigorous as traditional public key algorithms because it has a shorter history than algorithms like RSA. With increasing computing power, the length of keys is becoming a less important issue for PC-based applications.

PORTS AND PROTOCOLS[59]

A port is a logical connection. When using the Internet communications protocol, Transmission Control Protocol/Internet Protocol (TCP/IP), designating a port is the way a client program specifies a particular server program on a computer in a network. Basically, a port number is a way to identify the specific process to which an Internet or other network message is to be forwarded when it arrives at a server. For TCP, User Datagram Protocol (UDP) and Internet Control Message Protocol (ICMP), a port number is a 16-bit integer that is put in the header attached to a unit of information (a message unit). This port number is passed logically between client and server transport layers and physically between the transport layer and the Internet protocol layer and then forwarded.

---

[59] Moody, R. "Ports and Port Scanning: An Introduction," *ISACA Journal*, Volume 4, 2001

Higher-level applications that use TCP/IP such as the web protocol and HTTP use ports with preassigned numbers. These are well-known ports, to which numbers have been assigned by the Internet Assigned Numbers Authority (IANA). Some application processes are given port numbers dynamically when each connection is made.

PORT NUMBERS

Allowable port numbers range from 0 to 65535. Ports 0 to 1023 are reserved for certain privileged services—the well-known ports. For example, for the HTTP service, port 80 is defined as the default. Because it is preassigned, port 80 and some other ports do not have to be specified in the uniform resource locator (URL). That lets the user simply type in an Internet address or URL, such as www.isaca.org, without specifying the port number at the end of the URL; www.isaca.org:80, in this case. Either format will work in the browser.

Protocol Numbers and Assignment Services

Port numbers are divided into three ranges: the well-known ports, the registered ports and the dynamic and/or private ports. IANA records list all well-known and registered port numbers:
- **The well-known ports**—0 through 1023: Controlled and assigned by the IANA and, on most systems, can be used only by system (or root) processes or by programs executed by privileged users. The assigned ports use the first portion of the possible port numbers. Initially, these assigned ports were in the range 0 to 255. Currently, the range for assigned ports managed by the IANA has been expanded to the range 0 to 1023.
- **The registered ports**—1024 through 49151: Listed by the IANA and, on most systems, can be used by ordinary user processes or programs executed by ordinary users.
- **The dynamic and/or private ports**—49152 through 65535: Not listed by IANA because of their dynamic nature.

When a server program is started via a port connection, it is said to bind to its designated port number. When another client program wants to use that server, it also must send a request to bind to the designated port number. Ports are used in TCP to name the ends of logical connections that carry long-term conversations. To the extent possible, these same port assignments are used with UDP.

For example, a request for a file is sent through the browser software to a server accessible from the Internet. The request might be served from that host's File Transfer Protocol (FTP) application residing on a particular internal server. To pass the request to the FTP process residing on that remote server, the TCP software layer in the computer (the browser) specifies port number 21 (the IANA assigned port for an FTP request) in the 16-bit port number integer that is appended to the request as part of header information. At the remote server, the TCP layer will read the port number (21) and forward the request to the FTP program on the server. Common services are implemented on the same port across different platforms. For example, the service generally runs on port 80, whether using UNIX or Windows operating system. These transport layer mechanisms along with the IP addresses for a connection (sender and receiver) uniquely identify a connection.

In a basic Internet type of computer configuration using multiple computers, a site server is installed behind a firewall and its databases are installed on a second computer behind a second firewall. Other configurations are possible, of course. In some cases, there is a corporate intranet with internal users and a database server behind a firewall, a site server, another firewall and external users (it is also useful to have external site server tools access). Each firewall would allow access to ports set open by the site server software. With many server software applications, a number of ports are set by default, for example, HTTP-80, HTTPS-443 (the standard secure web server port number), SMTP-25 and others.

Tunneling

The use of a protocol to submit information for something other than the purpose of that protocol is called tunneling. In tunneling, malicious insiders or outside hackers use the protocol as an established pathway, or tunnel, directing the exchange of information for malicious purpose.

An ICMP tunnel uses ICMP echo requests and reply packets to establish a covert connection between two remote computers (a client and proxy.). ICMP tunneling can be used to bypass firewalls rules through obfuscation of the actual traffic. Depending on the implementation of the ICMP tunneling software, this type of connection can also be categorized as an encrypted communication channel between two computers. Without proper deep packet inspection or log review, network administrators will not be able to detect this type of traffic through their network.

HTTP tunneling is a technique by which communications performed using various network protocols are encapsulated using the HTTP protocol, the network protocols in question usually belonging to the TCP/IP family of protocols. The HTTP protocol, therefore, acts as a wrapper for a channel that the network protocol being tunneled uses to communicate. The HTTP stream with its covert channel is termed an HTTP tunnel.

HTTP tunnel software consists of client-server HTTP tunneling applications that integrate with existing application software, permitting them to be used in conditions of restricted network connectivity including firewalled networks, networks behind proxy servers, and network address translation.

VIRTUAL PRIVATE NETWORKS

When designing a VPN, it is important to ensure that the VPN can carry all types of data in a secure and private manner over any type of connection. Tunneling transports higher-layer data over a VPN by Layer 2 protocols. One end of the tunnel is the client, and the other end is a connectivity device or a remote access server. Common types of tunneling include:

- **Point-to-point tunneling protocol (PPTP)**—A Layer 2 protocol developed by Microsoft® that encapsulates point-to-point protocol data. It is simple, but less secure than other tunneling protocols.
- **Layer 2 tunneling protocol (L2TP)**—A protocol that encapsulates point-to-point protocol data and is compatible among different manufacturers' equipment. The end points do not have to reside on the same packet-switched network and can remain isolated from other traffic.
- **Secure Sockets Layer VPN**—A form of Layer 3 VPN that can be used with a standard Web browser and uses transport layer security (TLS) protocols to encrypt traffic.
- **IPSec VPN**—IPSec VPNs protect Layer 2 and 3 IP packets between remote networks or hosts and an IPSec gateway/node located at the edge of a private network.

VOICE-OVER INTERNET PROTOCOL (VOIP)[60]

Users often expect that all voice communications are confidential. Any VoIP device is an IP device; therefore, it is vulnerable to the same types of attacks as any other IP device. A hacker or virus could potentially bring down the data and voice networks simultaneously in a single attack. Also, VoIP networks are still vulnerable to sniffing, DoS, traffic-flow disruption and toll fraud. Sniffing would allow the disclosure of sensitive information (such as user information), resulting in identity theft, which may be used to attack other data subsystems. Port scanning is often a precursor to a potential sniffing of the VoIP network. The ability to network sniff is becoming easier as many tools are readily available from open source websites as opposed to highly expensive specialty diagnostic equipment used for time division multiplexing (TDM).

DoS, or the flooding of the data network with data, is a common issue in the protection of data networks but needs to be revisited as quality of service (QoS) becomes implemented for VoIP networks. The IP end point is often overlooked, but it can be singled out as a point of attack and flooded with data, causing the device to reboot and eventually become unusable.

Traffic flow disruption allows further exploitation of the previous two vulnerabilities, whereas the redirecting of packets facilitates the determination of packet routes, increasing the likelihood of sniffing.

Voice packets travel "in the clear" over IP networks, so they may be vulnerable to unauthorized sniffing. Unless network-based encryption is used, all voice RTP packets travel in the clear over the network and could be captured or copied by any network-monitoring device.

VoIP networks have a number of characteristics that make for special security requirements. There is no such thing as scheduled downtime in telephony. Outages may result in massive, widespread customer panic or outrage. There could also be disclosure of confidential information, which, like the loss of other kinds of data, could adversely affect the organization. Many security teams spend most of their time preventing outside attackers from penetrating

---

[60] Khan, K., "Introduction to Voice-over IP Technology", *ISACA Journal,* Volume 2, 2005, *www.isaca.org/Journal/Past-Issues/2005/Volume-2/Pages/Introduction-to-Voice-over-IP-Technology1.aspx*

a corporate firewall or Internet-accessible bastion servers. However, many companies spend little or no effort protecting the internal network infrastructure or servers from inside attacks. In the context of voice communications, a prime example is an employee listening to another employee's personal or company-confidential phone calls.

REMOTE ACCESS[61]

Remote access connectivity to their information resources is required for many organizations for different types of users, such as employees, vendors, consultants, business partners and customer representatives. In providing this capability, a variety of methods and procedures are available to satisfy an organization's business need for this level of access.

TCP/IP Internet-based remote access is a cost-effective approach that enables organizations to take advantage of the public network infrastructures and connectivity options available, under which Internet service providers manage modems and dial-in servers, and DSL and cable modems reduce costs further to an organization. To effectively use this option, organizations establish a virtual private network over the Internet to securely communicate data packets over this public infrastructure. Available VPN technologies apply the Internet Engineering Task Force (IETF) IPSec standard. Advantages are their ubiquity, ease of use, inexpensive connectivity, and read, inquiry or copy only access. Disadvantages include that they are significantly less reliable than dedicated circuits, lack a central authority, and can be difficult to troubleshoot.

Organizations should be aware that using VPNs to allow remote access to their systems can create holes in their security infrastructure. The encrypted traffic can hide unauthorized actions or malicious software that can be transmitted through such channels. Intrusion detection systems (IDSs) and virus scanners able to decrypt the traffic for analysis and then encrypt and forward it to the VPN end point should be considered as preventive controls. A good practice will terminate all VPNs to the same end point in a so called VPN concentrator and will not accept VPNs directed at other parts of the network. To reduce potential impacts of VPN access risk, architectural controls are implemented to restrict remote access traffic to selected security hardened/virus-protected systems, remote access portals and nonsensitive network segments.

Remote access risk includes:
- DoS, where remote users may not be able to gain access to data or applications that are vital for them to carry out their day-to-day business
- Malicious third parties, who may gain access to critical applications or sensitive data by exploiting weaknesses in communications software and network protocols
- Misconfigured communications software, which may result in unauthorized access or modification of an organization's information resources
- Misconfigured devices on the corporate computing infrastructure
- Host systems not secured appropriately, which could be exploited by an intruder gaining access remotely
- Physical security issues over remote users' computers

Remote access controls include:
- Policies and standards
- Proper authorizations
- Identification and authentication mechanisms
- Encryption tools and techniques, such as use of a VPN
- System and network management (e.g., NAC)
- Restrict access to controlled systems, networks and applications

---

[61] ISACA, *CISA Review Manual 26th Edition*, USA, 2015

TOPIC 5—OPERATING SYSTEM SECURITY

SYSTEM/PLATFORM HARDENING
System hardening is the process of implementing security controls on a computer system. It is common for most computer vendors to set the default controls to be open, allowing ease of use over security. This introduces significant vulnerabilities unless the system is hardened.

The process of determining what is hardened and to what level varies based on the operating system, installed applications, system/platform use and exposure. Also, the exact controls available for hardening vary. Some common controls include:
- Authentication and authorization
- File system permissions
- Access privileges
- Logging and system monitoring
- System services
- Configuration restrictions

Regardless of the specific operating system, system hardening should implement the principle of least privilege or access control.

MODES OF OPERATIONS
Most operating systems have two modes of operations—**kernel mode** for execution of privileged instructions for the internal operation of the system and **user mode** for normal activities. In kernel mode, there are no protections from errors or malicious activity and all parts of the system and memory are accessible. See **figure 4.9**.

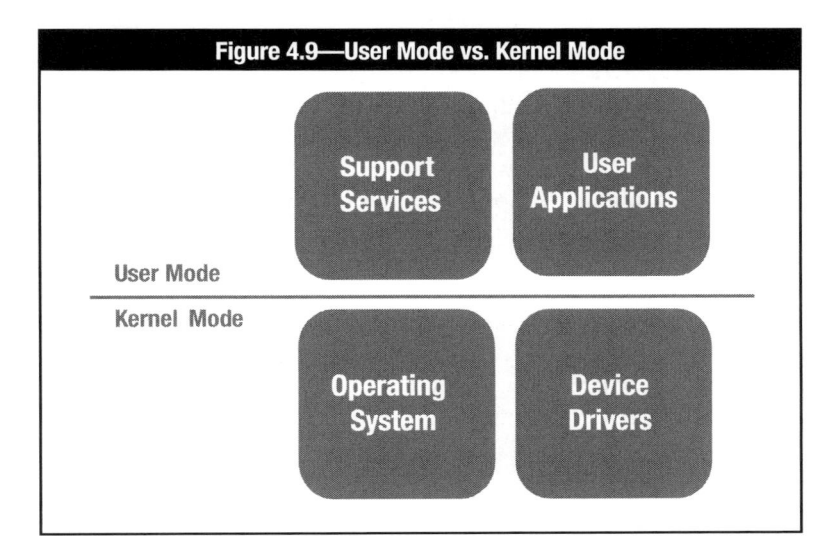

Figure 4.9—User Mode vs. Kernel Mode

Operating systems allow controlled access to kernel mode operations through system calls that usually require privileges. These privileges are defined on a user or program basis and should be limited under the principle of least privilege.

Most attacks seek to gain privileged or kernel mode access to the system in order to circumvent other security controls.

FILE SYSTEM PERMISSIONS
Operating systems have file systems that manage data files stored within the system and provide access controls to determine which users (or programs) have what type of access to a file. Common file accesses include creation, modification, read, write and deletion controls.

Section 4: Security of Networks, Systems, Applications and Data

CREDENTIALS AND PRIVILEGES
The access any particular user has to a system is controlled through a series of mechanisms. A user's credentials define who they are and what permissions they have to access resources within the system.

Passwords are the standard mechanism to authenticate a user to the system and must be managed correctly to ensure they are not easily guessed or compromised. Most operating systems provide controls around passwords such as minimum length, lifetime for any particular password and how many attempts to use a password are allowed before denying access.

Another key user control is the privileges assigned to a particular user. These privileges must be carefully chosen and controlled to prevent misuse or compromise. Assignment of privileges should follow the principle of least privilege required for a user to do their job.

Administrators can also limit the ways in which users can access systems. For example, administrators can set logon constraints based on the time of day, the total time logged on, the source address and unsuccessful logon attempts.

PLATFORM HARDENING
Security practitioners must understand the types and roles of accounts on each platform they are protecting. For example, Windows differentiates between files and devices such as printers, whereas everything in UNIX is considered to be a file to include physical devices.

Hardening is a process that reduces vulnerability by limiting the attack vectors that might be used as points of compromise. A hardened system is one that does not store any sensitive data that are not immediately needed to support a business operation. In addition, it has all unnecessary functionality disabled, including ports, services and protocols that are not required for the intended use of the system within the enterprise environment. Many devices and systems come with guest accounts or default passwords that should be changed or disabled as part of the hardening process.[62]

Server hardening is the hardening process, but applied to servers. Server hardening is important to protect an organization's servers from attack.

For the cybersecurity practitioner, identifying the location of critical information is imperative not only to security, but also incident response.

In UNIX, the following directories require additional consideration:
- /etc/passwd—Maintains user account and password information
- /etc/shadow—Retains the encrypted password of the corresponding account
- /etc/group—Contains group information for each account
- /etc/gshadow—Contains secure group account information
- /bin—Location of executable files
- /boot—Contains files for booting system
- /kernel—Kernel files
- /sbin—Contains executables, often for administration
- /usr—Include administrative commands

---

[62] ISACA, *CISA Review Manual 26th Edition*, USA, 2015

Section 4: Security of Networks, Systems, Applications and Data

For Windows, you need not look any further than the Registry—a central hierarchical database that stores configuration settings and options.[63] A hive is a logical group of keys, subkeys and values in the registry that has a set of supporting files and backups of its data.[64]
- HKEY_CURRENT_CONFIG—Contains volatile information generated at boot
- HKEY_CURRENT_USER—Settings specific to current user
- HKEY_LOCAL_MACHINE\SAM—Holds local and domain account information
- HKEY_LOCAL_MACHINE\Security—Contains security policy referenced and enforced by kernel
- HKEY_LOCAL_MACHINE\Software—Contains software and Windows settings
- HKEY_LOCAL_MACHINE\System—Contains information about Windows system setup
- HKEY_USERS\.DEFAULT—Profile for Local System account

Most of the supporting files for the hives are in the %SystemRoot%\System32\Config directory. These files are updated each time a user logs on.[65]

COMMAND LINE KNOWLEDGE

Cybersecurity professionals often use command line tools as part of their security routine. The following list provides 10 popular command line tools for cybersecurity:
- **Nmap**—Network port scanner and service detector
- **Metasploit**—Penetration testing software
- **Aircrack-ng**—802.11 WEP and WPA-PSK keys cracking program
- **Snort®**—Open-source IDS/IPS
- **Netstat**—Displays detailed network status information
- **Netcat**—Networking utility that reads and writes data across network connections, using the TCP/IP protocol
- **Tcpdump**—Command line packet analyzer
- **John the Ripper**—Password cracker
- **Kismet**—02.11 Layer 2 wireless network detector, sniffer and IDS
- **OpenSSH/PuTTY/SSH**—Program for logging into or executing commands on a remote machine. Useful UNIX commands for the cybersecurity practitioner are listed in **figure 4.10**.

Figure 4.10—UNIX Commands

Command	Description
finger {userid}	Display information about a user
cat	Display or concatenate file
cd	Change directory
chmod	Change file permissions Note: UNIX permissions are managed using octal notation by user, group, and others. Manipulating permissions is above the purpose of this material but is critical as you further your cybersecurity career.
cp	Copy
date	Display current date and time
diff	Display differences between text files
grep	Find string in file

---

[63] Microsoft Press, *Microsoft Computer Dictionary*, 5th edition, USA, 2002
[64] Microsoft, *Registry Hives*, http://msdn.microsoft.com/en-us/library/windows/desktop/ms724877%28v=vs.85%29.aspx
[65] *Ibid*

Section 4: Security of Networks, Systems, Applications and Data

Figure 4.10—UNIX Commands (cont.)

Command	Description
ls	Directory list. Useful switches: -a Display all files * -d Display only directories -l Display long listing -u Display files by access (newest first) -U Display results by creation (newest first) Note: Unlike Windows, UNIX does not afford the opportunity to "turn on" hidden files. Referred to as dot files, these file names begin with a ".", hence the name. To view these protected system files, you must use the –a switch. [ls –a or ls –al]
man	Displays help
mkdir	Make directory
mv	Move/rename file
ps	Display active processes
pwd	Displays the current directory
rm	Delete file
rmdir	Delete directory
sort	Sort data
whoami	Tells you who you are logged in as

VIRTUALIZATION

Virtualization provides an enterprise with a significant opportunity to increase efficiency and decrease costs in its IT operations.

At a high level, virtualization allows multiple OSs (guests), to coexist on the same physical server (host), in isolation of one another. Virtualization creates a layer between the hardware and the guest OSs to manage shared processing and memory resources on the host. Often, a management console provides administrative access to manage the virtualized system. There are advantages and disadvantages, as shown in **figure 4.11**.

Figure 4.11—Advantages and Disadvantages of Virtualization

Advantages	Disadvantages
• Server hardware costs may decrease for both server builds and server maintenance. • Multiple OSs can share processing capacity and storage space that often goes to waste in traditional servers, thereby reducing operating costs. • The physical footprint of servers may decrease within the data center. • A single host can have multiple versions of the same OS, or even different OSs, to facilitate testing of applications for performance differences. • Creation of duplicate copies of guests in alternate locations can support business continuity efforts. • Application support personnel can have multiple versions of the same OS, or even different OSs, on a single host to more easily support users operating in different environments. • A single machine can house a multitier network in an educational lab environment without costly reconfigurations of physical equipment. • Smaller organizations that had performed tests in the production environment may be better able to set up logically separate, cost-effective development and test environments. • If set up correctly, a well-built, single access control on the host can provide tighter control for the host's multiple guests.	• Inadequate configuration of the host could create vulnerabilities that affect not only the host, but also the guests. • Exploits of vulnerabilities within the host's configuration, or a DoS attack against the host, could affect all of the host's guests. • A compromise of the management console could grant unapproved administrative access to the host's guests. • Performance issues of the host's own OS could impact each of the host's guests. • Data could leak between guests if memory is not released and allocated by the host in a controlled manner. • Insecure protocols for remote access to the management console and guests could result in exposure of administrative credentials.

Source: ISACA, *CISA Review Manual 26th Edition*, USA, 2015, figure 5.14

Although virtualization offers significant advantages, those advantages come with risk that an enterprise must manage effectively. Because the host in a virtualized environment represents a potential single point of failure within the system, a successful attack on the host could result in a compromise that is larger in both scope and impact.

To address this risk, an enterprise can often implement and adapt the same principles and best practices for a virtualized server environment that it would use for a server farm. These include the following:
- Strong physical and logical access controls, especially over the host and its management console
- Sound configuration management practices and system hardening for the host, including patching, antivirus, limited services, logging, appropriate permissions and other configuration settings
- Appropriate network segregation, including the avoidance of virtual machines in the demilitarized zone (DMZ) and the placement of management tools on a separate network segment
- Strong change management practices

SPECIALIZED SYSTEMS

Some computer systems and applications are very specialized and may have unique threats and risk and require different types of controls.

Examples of specialized systems include **supervisory control and data acquisition (SCADA)** systems, **Industrial Control Systems (ICS)**, or other real-time monitoring or control systems that operate in specialized environments.

SCADA systems control industrial and manufacturing processes, power generation, air traffic control systems, and emergency communications and defense systems.

Historically, these systems were designed as stand-alone systems and because of the real-time nature of their applications often did not have any "overhead" software that would slow down operations. However, these systems are not commonly networked and often have few of the common controls found in more commercial systems.

Because of the importance of these systems on critical operations, they can be targeted by many different adversaries, and the impact of a successful attack can be catastrophic or even life threatening.

Many existing SCADA systems did not consider security in their design or deployment, and while vendors are improving security, these systems require careful assessment of risk and threats and often require special controls to compensate for inherent weaknesses.

TOPIC 6—APPLICATION SECURITY

Insecure applications open your organization up to external attackers who may try to use unauthorized code to manipulate the application to access, steal, modify or delete sensitive data. Application security measures should be applied during the design and development phase of the application, followed by routine security countermeasures used throughout the life cycle.

SYSTEM DEVELOPMENT LIFE CYCLE (SDLC)

Organizations often commit significant resources (e.g., people, applications, facilities and technology) to develop, acquire, integrate and maintain application systems that are critical to the effective functioning of key business processes. The SDLC process, shown in **figure 4.12**, guides the phases deployed in the development or acquisition of a software system and, depending on the methodology, may even include the controlled retirement of the system.

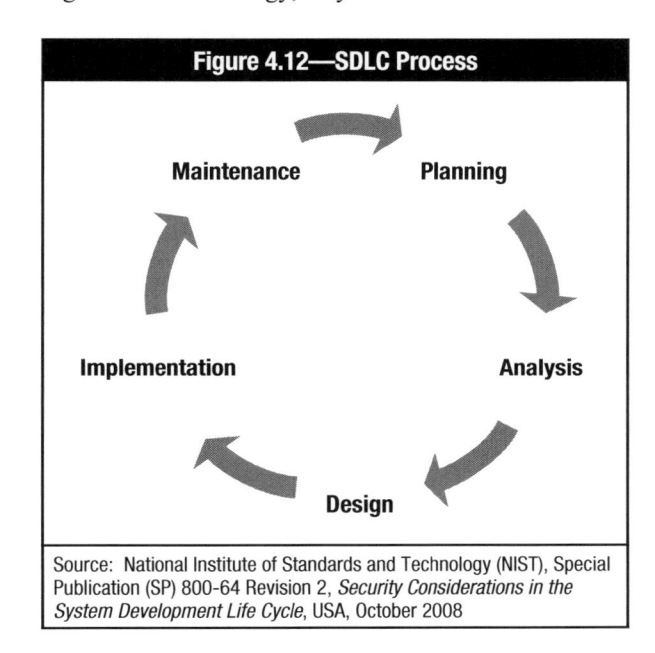

Figure 4.12—SDLC Process

Maintenance → Planning → Analysis → Design → Implementation → Maintenance

Source: National Institute of Standards and Technology (NIST), Special Publication (SP) 800-64 Revision 2, *Security Considerations in the System Development Life Cycle*, USA, October 2008

The SDLC includes:
- IT processes for managing and controlling project activity
- An objective for each phase of the life cycle that is typically described with key deliverables, a description of recommended tasks and a summary of related control objectives for effective management
- Incremental steps or deliverables that lay the foundation for the next phase

Specifically, the SDLC is a formal process to characterize design requirements and should include:
- Business requirements containing descriptions of what a system should do
- Functional requirements and the use of case models describing how users will interact with a system
- Technical requirements, design specifications and coding specifications describing how the system will interact, conditions under which the system will operate and the information criteria that the system should meet
- Risk mitigation and control requirements to protect the integrity of the system, confidentiality of information stored, processed or communicated as well as adequate authentication and authorization mechanisms

Security Within SDLC

The design and deployment of controls will often be undertaken as a systems development project. While there are several project management techniques that can be used to manage system development projects, they should be an integral and equal part of any SDLC process.

Section 4: Security of Networks, Systems, Applications and Data

Design Requirements

Not considering the security in the design of a system or application is one of the major contributing factors to today's cybersecurity vulnerabilities, making it easier for systems to be compromised. Too often, security is an afterthought, and controls are retrofitted in an *ad hoc* way only after security weaknesses are identified.

Security and risk mitigation should be formal design criteria in any SDLC process and start with threat and risk assessment of the proposed system, identification of controls, implementation of those controls, and testing and review.

OWASP Top Ten

The Open Web Application Security Project (OWASP) publishes a list of the top 10 application security risks. **Figure 4.13** provides the top 10 application security risks for 2013.

Figure 4.13—Top 10 Application Security Risks in 2013	
Attack Vector	**Description of Security Risk**
Injection	Injection flaws occur when untrusted data is sent to an interpreter. The attacker can trick the interpreter into executing unintended commands or accessing unauthorized data. Injection flaws are prevalent and are often found in SQL and LDAP queries and OS commands.
Broken Authentication and Session Management	If an application function related to authentication or session management is not implemented correctly, it can allow an attacker to compromise passwords, keys or session tokens and impersonate users.
Cross-Site Scripting (XSS)	XSS flaws occur when an application takes untrusted data and sends it to a web browser without proper validation. This is the most prevalent web application security flaw. Attackers can use XSS to hijack user sessions, insert hostile content, deface websites and redirect users.
Insecure Direct Object References	A direct object reference occurs when a developer exposes a reference to an internal implementation object. Attackers can manipulate these references to access unauthorized data.
Security Misconfiguration	Security settings must be defined, implemented and maintained for applications, frameworks, application servers, web servers, database servers and platforms. Security misconfiguration can give attackers unauthorized access to system data or functionality.
Sensitive Data Exposure	If web applications do not properly secure sensitive data through the use of encryption, attackers may steal or modify sensitive data such as health records, credit cards, tax IDs and authentication credentials.
Missing Function Level Access Control	When function level access rights are not verified, attackers can forge requests to access functionality without authorization.
Cross-Site Request Forgery (CSRF)	A CSRF attack occurs when an attacker forces a user's browser to send forged HTTP requests, including session cookies. This allows an attacker to trick victims into performing operations on the illegitimate website.
Using Components with Known Vulnerabilities	Certain components such as libraries, frameworks and other software modules usually run with full privileges. Attackers can exploit a vulnerable component to access data or take over a server.
Unvalidated Redirects and Forwards	Web applications frequently redirect or forward users to other pages. When untrusted data are used to determine the destination, an attacker can redirect victims to phishing or malware sites.

Application controls are controls over input, processing and output functions. They include methods to help ensure data accuracy, completeness, validity, verifiability and consistency, thus achieving data integrity and data reliability.

Application controls may consist of edit tests; totals; reconciliations and identification; and reporting of incorrect, missing or exception data. Automated controls should be coupled with manual procedures to ensure proper investigation of exceptions. Implementation of these controls helps ensure system integrity; that applicable system functions operate as intended; and that information contained by the system is relevant, reliable, secure and available when needed. Application controls include:
- Firewalls
- Encryption programs
- Anti-malware programs
- Spyware detection/removal programs
- Biometric authentication

To reduce application security risk, OWASP recommends the following:
- Define application security requirements.
- Use good application security architecture practices from the start of the application design.
- Build strong and usable security controls.
- Integrate security into the development life cycle.
- Stay current on application vulnerabilities.

Testing
The testing phase of SDLC includes:
- Verification and validation that a program, subsystem or application and the designed security controls perform the functions for which they have been designed
- Determination of whether the units being tested operate without any malfunction or adverse effect on other components of the system
- A variety of development methodologies and organizational requirements to provide for a large range of testing schemes or levels

From a security perspective, this should include vulnerability and control testing.

Review Process
Code review processes vary from informal processes to very formal walk-throughs, team review or code inspections. Security should be an integrated part of any review process.

SEPARATION OF DEVELOPMENT, TESTING AND PRODUCTION ENVIRONMENTS
Development and testing environments are relatively open and often have fewer access controls due to the collaborative nature of the development process. It is important to separate the development, testing and production environments to minimize a compromise or misconfiguration being introduced or cascading through the process. Different access controls (credentials) should be used between the different environments.

Also, if production data are used in the test environment, private or personally identifiable information should be scrambled so that confidential information is not inadvertently disclosed.

Agile Development
Agile allows for projects, including software development, to be built in a more flexible, iterative fashion in order to respond more quickly to changes that occur during a project. Many benefits to using agile as opposed to the traditional waterfall method have been reported; however, it has not been regularly adopted in the security realm. Rather than working penetration testing and other security measures into earlier stages of the process, these often occur close to the release of a project, making it difficult to make changes in response to the test results. This can result in workarounds and quick fixes being developed to meet a release deadline as opposed to more thoughtful, secure solutions.

Agile, in contrast, works opportunities for reevaluation of the project within the project plan itself and often includes many opportunities for early testing. This approach allows for the schedule to adapt to testing findings in order to develop more thoughtful solutions, rather than rushing to fix an unknown to make a release date.

Development and IT Operations (DevOps)[66]
DevOps combines the concepts of agile development, agile infrastructure and flexible operations to enable rapid and continuous releases and ongoing improvement in IT value creation. The DevOps movement was built out from frustration of IT groups with the dysfunction and deficient tools and processes, with the aim of making software development and operations more efficient and less painful.

---

[66] ISACA, *DevOps Overview*, USA, 2015, www.isaca.org/dev-ops

DevOps breaks large projects into smaller deliverables and multiple deployments, which are easier to manage from design to deployment and operations. Iterative and frequent deployments can be orchestrated to move seamlessly from one group to the next, until they are promoted to production with minimal risk of disruptions. Small deployments are easier to debug along the development process and are stabilized after they are in operation.

Some business performance benefits of DevOps include:
- Reduced time to market
- Faster return on investment
- High performance
- Increased quality
- Customer satisfaction
- Reduced IT waste
- Improved supplier and business partner performance
- Reduction to the human factor threat

Challenges to using DevOps include:
- Misconception about what DevOps means
- Belief that DevOps is not concerned with compliance and security
- Need for automation
- Lack of skills
- Organizational culture
- Fear of change
- Silo mentality

ADDITIONAL THREATS

It is important for the aspiring security practitioner to recognize that there are many sources of security-related advice, best practices and recommendations. Just because a threat does not make it into a "top" list for a year does not mean that you can forget it.

Other security threats to be aware of include:
- **Covert channel**—Means of illicitly transferring information between systems using existing infrastructure. Covert channels are simple, stealthy attacks that often go undetected.
- **Race condition**—According to Rouse, "an undesirable situation that occurs when a device or system attempts to perform two or more operations at the same time, but because of the nature of the device or system, the operations must be done in the proper sequence in order to be done correctly."[67] Race conditions vary; however, these vulnerabilities all afford opportunities for unauthorized network access.
- **Return-oriented programming attack**—Frequently used technique to exploit memory corruption vulnerabilities. Simply stated, it allows an attacker to execute code despite the technological advances such as nonexecutable stacks and nonexecutable heaps. Memory corruption vulnerabilities occur "when a privileged program is coerced into corrupting its own memory space, such that the memory areas corrupted have an impact on the secure functioning of the program."[68]
- **Steganography**—The art or practice of concealing a message, image or file within another message, image or file. Media files are ideal because of their large size.

---

[67] Rouse, Margaret, "Race Condition," September 2005, http://searchstorage.techtarget.com/definition/race-condition
[68] Herath, Nishad, "The State of Return Oriented Programming in Contemporary Exploits," Security Intelligence, 3 March 2014, http://securityintelligence.com/return-oriented-programming-rop-contemporary-exploits/#.VFkNEBa9bD0

TOPIC 7—DATA SECURITY

Databases can be individually protected with control that is similar to protections applied at the system level. Specific controls that can be placed at the database level include:
- Authentication and authorization of access
- Access controls limiting or controlling the type of data that can be accessed and what types of accesses are allowed (such as read-only, read and write, or delete)
- Logging and other transactional monitoring
- Encryption and integrity controls
- Backups

The controls used to protect databases should be designed in conjunction with system and application controls and form another layer of protection in a defense in depth scheme.

DATA CLASSIFICATION

The information an organization uses can be of varying value and importance. For example, some information may be public and require minimal protection while other information such as national security information, health or other personal information, or trade secrets could result in significant harm to the organization if inadvertently released, deleted or modified.

It is important for an organization to understand the sensitivity of information and classify data based on its sensitivity and the impact of release or loss of the information.

Data classification works by tagging data with metadata based on a classification taxonomy. This enables data to be found quickly and efficiently, cuts back on storage and backup costs, and helps to allocate and maximize resources. Classification levels should be kept to a minimum. They should be simple designations that assign different degrees of sensitivity and criticality.

Data classification should be defined in a data classification policy that provides definition of different classes of information and how each class of information should be handled and protected. In addition, the classification scheme should convey the association of the data and their supporting business processes.

In some cases, local regulations may impact data classification and handling such as those controlled by data protection acts. For example, the US Sarbanes-Oxley Act defines which data records must be stored and for how long.

Information may also need to be reclassified based on changes to its importance. For example, prior to a product release, details of the design, pricing and other information may be confidential and need significant protection; however, after the product is announced, this information may become public and not require the same levels of protection.

DATA OWNERS

Another important consideration for data security is defining the data owner. Although IT applies the security controls and monitoring of business data, the data do not belong to IT. Business information belongs to whoever is ultimately responsible for the business process. The owner is usually responsible for determining the data classification and, therefore, the level of protection required. The data owner may be an individual who creates the data or an organizational element that acts as a custodian of the information. The data classification process is shown in **figure 4.14**.

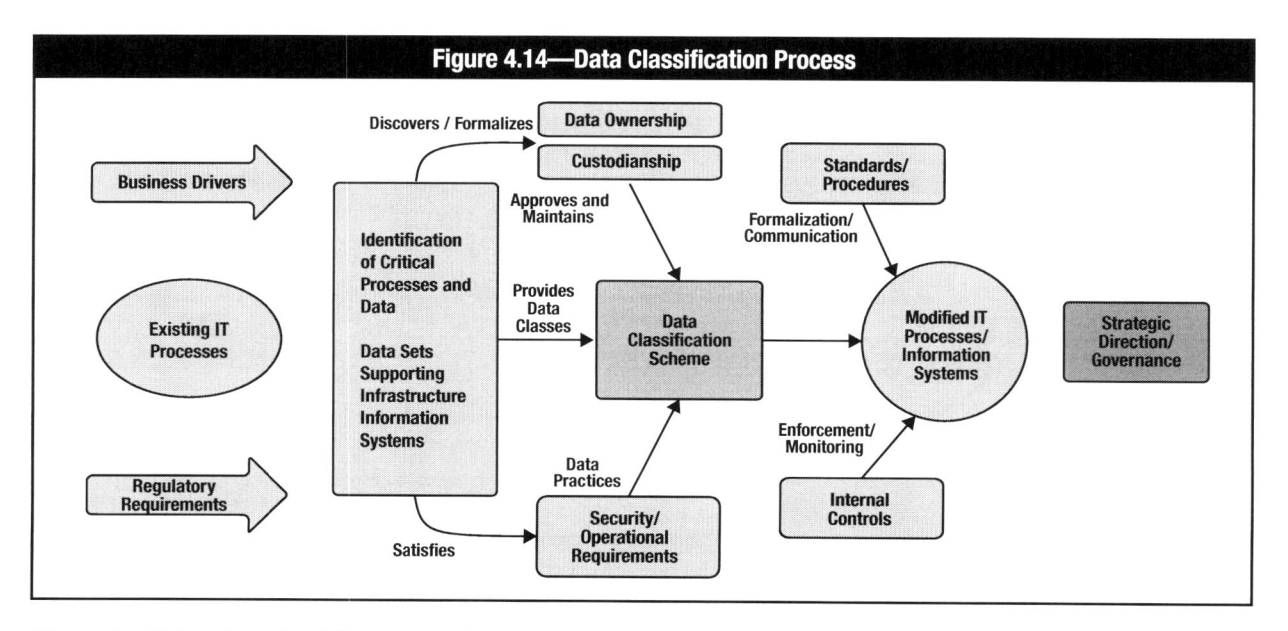

Figure 4.14—Data Classification Process

When classifying data, the following requirements should be considered:
- **Access and authentication**—Determine access requirements including defining users profiles, access approval criteria and validation procedures.
- **Confidentiality**—Determine where sensitive data are stored and how they are transmitted.
- **Privacy**—Use controls to warn an affected user that his or her information is about to be used.
- **Availability**—Determine the uptime and downtime tolerances for different data types.
- **Ownership and distribution**—Establish procedures to protect data from unauthorized copy and distribution.
- **Integrity**—Protect data from unauthorized changes using change control procedures and automated monitoring and detection for unauthorized changes and manipulation.
- **Data retention**—Determine retention periods and preserve specific versions of software, hardware, authentication credentials and encryption keys to ensure availability.
- **Auditability**—Keep track of access, authorizations, changes and transactions.

After data classification has been assigned, security controls can be established such as encryption, authentication and logging. Security measures should increase as the level of data sensitivity or criticality increases. The full data life cycle is shown in **figure 4.15**.

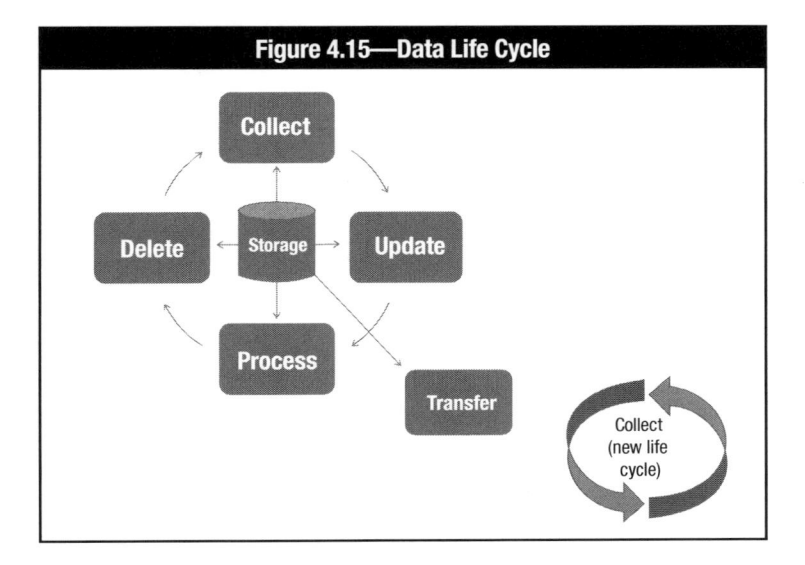

Figure 4.15—Data Life Cycle

DATABASE SECURITY

Database security covers a wide range of information security controls to protect the databases in an organization. Database security is essential because databases contain critical and sensitive information that is required to support business operations. Databases are vulnerable to types of risk, including:
- Unauthorized activity by authorized users
- Malware infections or interactions
- Capacity issues
- Physical damage
- Design flaws
- Data corruption

Database security is provided through the following:[69]
- Encryption of sensitive data in the database
- Use of database views to restrict information available to a user
- Secure protocols to communicate with the database
- Content-based access controls that restrict access to sensitive records
- Restricting administrator-level access
- Efficient indexing to enhance data retrieval
- Backups of databases (shadowing, mirroring)
- Backups of transaction journals (remote journaling)
- Referential integrity
- Entity integrity
- Validation of input
- Defined data fields (schema)
- Layered network access restrictions or segregation

[69] ISACA, *CISA Review Manual 26th Edition*, USA, 2015

Section 4: Security of Networks, Systems, Applications and Data

SECTION 4—KNOWLEDGE CHECK

1. Put the steps of the penetration testing phase into the correct order.
   A. Attack
   B. Discovery
   C. Reporting
   D. Planning

2. System hardening should implement the principle of _____ or _____ .
   A. Governance, compliance
   B. Least privilege, access control
   C. Stateful inspection, remote access
   D. Vulnerability assessment, risk mitigation

3. Select all that apply. Which of the following are considered functional areas of network management as defined by ISO?
   A. Accounting management
   B. Fault management
   C. Firewall management
   D. Performance management
   E. Security management

4. Virtualization involves:
   A. the creation of a layer between physical and logical access controls.
   B. multiple guests coexisting on the same server in isolation of one another.
   C. simultaneous use of kernel mode and user mode.
   D. DNS interrogation, WHOIS queries and network sniffing.

5. Vulnerability management begins with an understanding of cybersecurity assets and their locations, which can be accomplished by:
   A. vulnerability scanning.
   B. penetration testing.
   C. maintaining an asset inventory.
   D. using command line tools.

See answers in Appendix C.

Section 5: Incident Response

Topics covered in this section include:
1. Event vs. incident
2. Security incident response
3. Investigations, legal holds and preservation
4. Forensics
5. Disaster recovery and business continuity plans

Page intentionally left blank

TOPIC 1—EVENT VS. INCIDENT

All organizations need to put significant effort into protecting themselves and preventing cyberattacks from causing harm or disruption. However, security controls are not perfect and cannot completely eliminate all risk; therefore, it is important that organizations prepare for, and are capable of detecting and managing, potential cybersecurity problems.

EVENT VS. INCIDENT

It is important to distinguish between an event and an incident because the two terms are often used synonymously, even though they have different meanings. An event is any change, error or interruption within an IT infrastructure such as a system crash, a disk error or a user forgetting their password. The National Institute of Standards and Technology (NIST) defines an event as "any observable occurrence in a system or network."[70]

While there is general agreement on what an event is, there is a greater degree of variety in defining an incident. NIST defines an incident as "a violation or imminent threat of violation of computer security policies, acceptable use policies, or standard security practices." Another commonly used definition is "the attempted or successful unauthorized access, use, disclosure, modification or loss of information or interference with system or network operations." Many organizations define an incident as the activity of a human threat agent. Others would include anything disruptive, including a court order for discovery of electronic information or disruption from a natural disaster.

Regardless of the exact definition used by a particular organization, it is important to distinguish between events that are handled in the normal course of business and incidents that require security and investigative expertise to manage.

TYPES OF INCIDENTS

A cybersecurity incident is an adverse event that negatively impacts the confidentiality, integrity and availability of data. Cybersecurity incidents may be unintentional, such as someone forgetting to activate an access list in a router, or intentional, such as a targeted attack by a hacker. These events may also be classified as technical or physical. Technical incidents include viruses, malware, denial-of-service (DoS) and system failure. Physical incidents may include social engineering and lost or stolen laptops or mobile devices.

---

[70] National Institute of Standards and Technology (NIST), *Special Publication 800-61 Revision 2, Computer Security Incident Handling Guide*, USA, August 2012

Section 5: Incident Response

There are many types of cybersecurity-related incidents, and new types of incidents emerge frequently. US-CERT provides a common set of terms and their relationships developed from NIST SP 800-61 Revision 2, as seen in **figure 5.1.**

Figure 5.1—Attack Vectors Taxonomy		
Attack Vector	**Description**	**Example**
Unknown	Cause of attack is unidentified.	This option is acceptable if cause (vector) is unknown upon initial report. The attack vector may be updated in a follow-up report.
Attrition	An attack that employs brute force methods to compromise, degrade, or destroy systems, networks or services	Denial of service intended to impair or deny access to an application; a brute force attack against an authentication mechanism, such as passwords or digital signatures
Web	An attack executed from a website or web-based application.	Cross-site scripting attack used to steal credentials, or a redirect to a site that exploits a browser vulnerability and installs malware
Email/Phishing	An attack executed via an email message or attachment	Exploit code disguised as an attached document, or a link to a malicious website in the body of an email message
External/Removable Media	An attack executed from removable media or a peripheral device	Malicious code spreading onto a system from an infected flash drive
Impersonation/Spoofing	An attack involving replacement of legitimate content/services with a malicious substitute	Spoofing, man in the middle attacks, rogue wireless access points and structured query language injection attacks all involve impersonation.
Improper Usage	Any incident resulting from violation of an organization's acceptable usage policies by an authorized user, excluding the above categories	User installs file-sharing software, leading to the loss of sensitive data; or a user performs illegal activities on a system.
Loss or Theft of Equipment	The loss or theft of a computing device or media used by the organization	A misplaced laptop or mobile device
Other	An attack method does not fit into any other vector	
Source: US-CERT, "Attack Vectors Taxonomy," US-CERT Federal Incident Notification Guidelines, USA, https://www.us-cert.gov/incident-notification-guidelines		

The European Union Agency for Network and Information Security (ENISA) provides a taxonomy for incidents, shown in **figure 5.2**.

Figure 5.2—European CSIRT Network Taxonomy		
Incident Class (mandatory input field)	**Incident Type (optional but desired input field)**	**Description/Examples**
Abusive Content	Spam	Unsolicited bulk email, meaning that the recipient has not granted verifiable permission for the message to be sent and that the message is sent as part of a larger collection of messages, all having identical content
	Harmful speech	Discreditation or discrimination of somebody (e.g., cyber stalking, racism and threats against one or more individuals)
	Child/sexual/violence	Child pornography, glorification of violence, etc.
Malicious Code	Virus	Software that is intentionally included or inserted in a system for a harmful purpose. A user interaction is normally necessary to activate the code.
	Worm	
	Trojan	
	Spyware	
	Dialer	
	Rootkit	

Figure 5.2—European CSIRT Network Taxonomy (cont.)		
Incident Class (mandatory input field)	**Incident Type (optional but desired input field)**	**Description/Examples**
Information Gathering	Scanning	Attacks that send requests to a system to discover weak points. This also includes some kinds of testing processes to gather information about hosts, services and accounts. Examples: fingerd, DNS querying, ICMP, SMTP (EXPN, RCPT, etc.).
	Sniffing	Observing and recording network traffic (wiretapping)
	Social engineering	Gathering information from a human being in a nontechnical way (e.g., lies, tricks, bribes or threats)
Intrusion Attempts	Exploiting known vulnerabilities	An attempt to compromise a system or to disrupt any service by exploiting vulnerabilities with a standardised identifier such as CVE name (e.g., buffer overflow, backdoors, cross side scripting, etc.).
	Login attempts	Multiple login attempts (guessing/cracking of passwords, brute force)
	New attack signature	An attempt using an unknown exploit
Intrusions	Privileged account compromise	A successful compromise of a system or application (service). This can have been caused remotely by a known or new vulnerability, but also by an unauthorized local access. Also includes being part of a botnet.
	Unprivileged account compromise	
	Application compromise	
	Bot	
Availability	DoS	By this kind of an attack a system is bombarded with so many packets that the operations are delayed or the system crashes. DoS examples are ICMP and SYN floods, Teardrop attacks and mail-bombing. DDoS often is based on DoS attacks originating from botnets, but also other scenarios exist like DNS Amplification attacks.
	DDoS	
	Sabotage	
	Outage (no malice)	
Information Content Security	Unauthorized access to information	Besides local abuse of data and systems, the security of information can be endangered by successful compromise of an account or application. In addition, attacks that intercept and access information during transmission (wiretapping, spoofing or hijacking) are possible. Human/configuration/software error can also be the cause.
	Unauthorized modification of information	
Fraud	Unauthorized use of resources	Using resources for unauthorized purposes including profit-making ventures (e.g., the use of email to participate in illegal profit chain letters or pyramid schemes)
	Copyright	Selling or installing copies of unlicensed commercial software or other copyright protected materials (Warez)
	Masquerade	Types of attacks in which one entity illegitimately assumes the identity of another in order to benefit from it
	Phishing	Masquerading as another entity in order to persuade the user to reveal a private credential.
Vulnerable	Open for abuse	Open resolvers, world readable printers, vulnerability apparent from Nessus, etc. scans, virus, signatures not up to date, etc.
Other	All incidents which do not fit in one of the given categories should be put into this class.	If the number of incidents in this category increases, it is an indicator that the classification scheme must be revised.
Test	Meant for testing	Meant for testing.
Source: ENISA, *Existing taxonomies*, www.enisa.europa.eu/topics/csirt-cert-services/community-projects/existing-taxonomies		

Page intentionally left blank

TOPIC 2—SECURITY INCIDENT RESPONSE

WHAT IS INCIDENT RESPONSE?

Incident response is a formal program that prepares an entity for an incident. Incident response phases are shown in **figure 5.3**. Incident response generally includes:
1. **Preparation** to establish roles, responsibilities and plans for how an incident will be handled
2. **Detection and Analysis** capabilities to identify incidents as early as possible and effectively assess the nature of the incident
3. **Investigation** capability if identifying an adversary is required
4. **Mitigation and Recovery** procedures to contain the incident, reduce losses and return operations to normal
5. **Postincident Analysis** to determine corrective actions to prevent similar incidents in the future

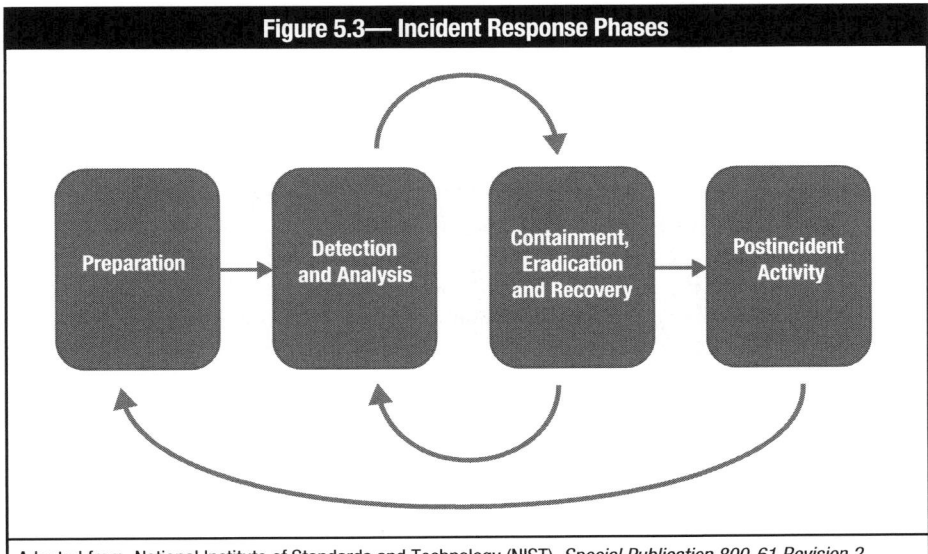

Figure 5.3— Incident Response Phases

Preparation → Detection and Analysis → Containment, Eradication and Recovery → Postincident Activity

Adapted from: National Institute of Standards and Technology (NIST), *Special Publication 800-61 Revision 2, Computer Security Incident Handling Guide*, Figure 3-1. Incident Response Life Cycle, USA, August 2012, http://nvlpubs.nist.gov/nistpubs/SpecialPublications/NIST.SP.800-61r2.pdf

WHY DO WE NEED INCIDENT RESPONSE?

Waiting until an incident occurs to figure out what to do is a recipe for disaster. Adequate incident response planning and implementation allows an organization to respond to an incident in a systematic manner that is more effective and timely. Organizations that do not plan for a cybersecurity incident will suffer greater losses for a more extended period of time. The current trend shows an increase in incident occurrences. These attacks are becoming more sophisticated and are resulting in escalating losses.

In addition, many national regulations and international standards require the development of incident response capabilities. Compliance regulations provide strict requirements for security policies and incident response planning.

ELEMENTS OF AN INCIDENT RESPONSE PLAN (IRP)[71]

A common approach to developing an IRP is a six-phase model of incident response including preparation, identification, containment, eradication, restoration and follow-up:
- **Preparation**—This phase prepares an organization to develop an IRP prior to an incident. Sufficient preparation facilitates smooth execution. Activities in this phase include:
  - Establishing an approach to handle incidents
  - Establishing policy and warning banners in information systems to deter intruders and allow information collection
  - Establishing communication plan to stakeholders

---

[71] ISACA, *CISM Review Manual 15th Edition*, USA, 2016

- Developing criteria on when to report an incident to authorities
- Developing a process to activate the incident management team
- Establishing a secure location to execute the IRP
- Ensuring equipment needed is available
- **Identification**—This phase aims to verify if an incident has happened and to find out more details about the incident. Reports on possible incidents may come from information systems, end users or other organizations. Not all reports are valid incidents, as they may be false alarms or may not qualify as an incident. Activities in this phase include:
  - Assigning ownership of an incident or potential incident to an incident handler
  - Verifying that reports or events qualify as an incident
  - Establishing chain of custody during identification when handling potential evidence
  - Determining the severity of an incident and escalating it as necessary
- **Containment**—After an incident has been identified and confirmed, the incident management team (IMT) is activated and information from the incident handler is shared. The team will conduct a detailed assessment and contact the system owner or business manager of the affected information systems/assets to coordinate further action. The action taken in this phase is to limit the exposure. Activities in this phase include:
  - Activating the IMT/IRT to contain the incident
  - Notifying appropriate stakeholders affected by the incident
  - Obtaining agreement on actions taken that may affect availability of a service or risk of the containment process
  - Getting the IT representative and relevant virtual team members involved to implement containment procedures
  - Obtaining and preserving evidence
  - Documenting and taking backups of actions from this phase onward
  - Controlling and managing communication to the public by the public relations team
- **Eradication**—When containment measures have been deployed, it is time to determine the root cause of the incident and eradicate it. Eradication can be done in a number of ways: restoring backups to achieve a clean state of the system, removing the root cause, improving defenses and performing vulnerability analysis to find further potential damage from the same root cause. Activities in this phase include:
  - Determining the signs and cause of incidents
  - Locating the most recent version of backups or alternative solutions
  - Removing the root cause. In the event of worm or virus infection, it can be removed by deploying appropriate patches and updated antivirus software.
  - Improving defenses by implementing protection techniques
  - Performing vulnerability analysis to find new vulnerabilities introduced by the root cause
- **Recovery**—This phase ensures that affected systems or services are restored to a condition specified in the service delivery objectives (SDO) or business continuity plan (BCP). The time constraint up to this phase is documented in the recovery time objective (RTO). Activities in this phase include:
  - Restoring operations as defined in the SDO
  - Validating that actions taken on restored systems were successful
  - Getting involvement of system owners to test the system
  - Facilitating system owners to declare normal operation
- **Lessons learned**—At the end of the incident response process, a report should always be developed to share what has happened, what measures were taken and the results after the plan was executed. Part of the report should contain lessons learned that provide the IMT and other stakeholders valuable learning points of what could have been done better. These lessons should be developed into a plan to enhance the incident management capability and the documentation of the IRP. Activities in this phase include:
  - Writing the incident report
  - Analyzing issues encountered during incident response efforts
  - Proposing improvement based on issues encountered
  - Presenting the report to relevant stakeholders

TOPIC 3—INVESTIGATIONS, LEGAL HOLDS AND PRESERVATION

INVESTIGATIONS

Cybersecurity incident investigations include the collection and analysis of evidence with the goal of identifying the perpetrator of an attack or unauthorized use or access. This may overlap with, but is distinctly separate from, the technical analysis used in incident response where the objective is to understand the nature of the attack, what happened and how it occurred.

The goals of an investigation can conflict with the goals of incident response. Investigations may require the attack or unauthorized access to continue while it is analyzed and evidence is collected, whereas remediation may destroy evidence or preclude further investigation. The organization's management must be an integral part of making decisions between investigating and remediation.

Investigations may be conducted for criminal activity (as defined by governmental statutes and legislation), violations of contracts or violations of an organization's policies. Cybersecurity investigators may also assist in other types of investigations where computers or networks were used in the commission of other crimes, such as harassment where email was used.

An investigation may take place entirely in-house or may be conducted by a combination of in-house personnel, service providers, and law enforcement or regulators.

EVIDENCE PRESERVATION

It is very important to preserve evidence in any situation. Most organizations are not well equipped to deal with intrusions and electronic crimes from an operational and procedural perspective, and they respond to it only when the intrusion has occurred and the risk is realized. The evidence loses its integrity and value in legal proceedings if it has not been preserved and subject to a documented chain of custody. This happens when the incident is inappropriately managed and responded to in an ad hoc manner.

The evidence of a computer crime exists in the form of log files, file time stamps, contents of memory, etc. Other sources include browser history, contact lists, cookies, documents, hidden files, images, metadata, temporary files and videos. While not comprehensive, it helps provide context for the cybersecurity professional as to how much information is available to responders. The ability to locate and capture evidence is dependent on data type, investigators' skills and experience, and tools.

Rebooting the system or accessing files could result in such evidence being lost, corrupted or overwritten. Therefore, one of the first steps taken should be copying one or more images of the attacked system. Memory content should also be dumped to a file before rebooting the system. Any further analysis must be performed on an image of the system and on copies of the memory dumped—not on the original system in question.

In addition to protecting the evidence, it is also important to preserve the chain of custody. **Chain of custody** is a term that refers to documenting, in detail, how evidence is handled and maintained, including its ownership, transfer and modification. This is necessary to satisfy legal requirements that mandate a high level of confidence regarding the integrity of evidence.

For evidence to be admissible in a court of law, the chain of custody needs to be maintained accurately and chronologically. The chain of evidence essentially contains information regarding:
- Who had access to the evidence (chronological manner)
- The procedures followed in working with the evidence (such as disk duplication, virtual memory dump)
- Proof that the analysis is based on copies that are identical to the original evidence (could be documentation, checksums, time stamps)

LEGAL REQUIREMENTS

Investigations have clearly defined legal requirements and these vary from country to country. Only trained investigators working with legal counsel should undertake investigations. Some of the legal issues that may be applicable include:
- Evidence collection and storage
- Chain of custody of evidence
- Searching or monitoring communications
- Interviews or interrogations
- Law enforcement involvement
- Labor, union and privacy regulation

These and other legal considerations are evolving when applied to cyberspace and vary, sometimes significantly, from jurisdiction to jurisdiction. Failure to perform an investigation in compliance with the appropriate legal requirements may create criminal or civil liabilities for the investigator and organization or may result in an inability to pursue legal remedies.

Many attacks are international in scope, and navigating the different (and sometimes conflicting) legal issues can be challenging, adding complexity to cybersecurity investigations. In some countries, private individuals and organizations are not permitted to carry out investigations and require law enforcement.

TOPIC 4—FORENSICS

By definition, digital forensics is the "process of identifying, preserving, analyzing and presenting digital evidence in a manner that is legally acceptable in any legal proceedings (i.e., a court of law)."[72] Computer forensics includes activities that involve the exploration and application of methods to gather, process, interpret and use digital evidence that help to substantiate whether an incident happened, such as:
- Providing validation that an attack actually occurred
- Gathering digital evidence that can later be used in judicial proceedings

Any electronic document or data can be used as digital evidence, provided there is sufficient manual or electronic proof that the contents of digital evidence are in their original state and have not been tampered with or modified during the process of collection and analysis.

It is important to use industry-specified best practices, proven tools and due diligence to provide reasonable assurance of the quality of evidence. It is also important to demonstrate integrity and reliability of evidence for it to be acceptable to law enforcement authorities. For example, if the IS auditor "boots" a computer suspected of containing stored information that might represent evidence in a court case, the auditor cannot later deny that they wrote data to the hard drive because the boot sequence writes a record to the drive. This is the reason specialist tools are used to take a true copy of the drive, which is then used in the investigation.

There are four major considerations in the chain of events in regards to evidence in digital forensics (**figure 5.4**):
- **Identify**—Refers to the identification of information that is available and might form the evidence of an incident.
- **Preserve**—Refers to the practice of retrieving identified information and preserving it as evidence. The practice generally includes the imaging of original media in presence of an independent third party. The process also requires being able to document chain-of-custody so that it can be established in a court of law.
- **Analyze**—Involves extracting, processing and interpreting the evidence. Extracted data could be unintelligible binary data after it has been processed and converted into human readable format. Interpreting the data requires an in-depth knowledge of how different pieces of evidence may fit together. The analysis should be performed using an image of media and not the original.
- **Present**—Involves a presentation to the various audiences such as management, attorneys, court, etc.

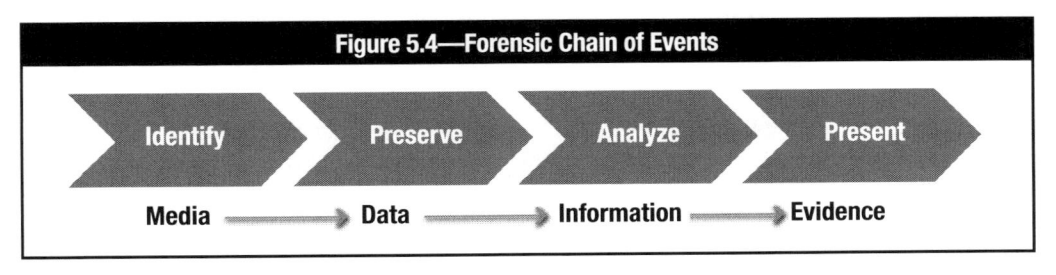

Figure 5.4—Forensic Chain of Events

Identify → Preserve → Analyze → Present

Media → Data → Information → Evidence

Acceptance of the evidence depends upon the manner of presentation (it should be convincing), qualifications of the presenter, and credibility of the process used to preserve and analyze the evidence. The assurance professional should give consideration to key elements of computer forensics during audit planning. These key elements are described in the following sections.

DATA PROTECTION

To prevent sought-after information from being altered, all measures must be in place. It is important to establish specific protocols to inform appropriate parties that electronic evidence will be sought and to not destroy it by any means. Infrastructure and processes for incident response and handling should be in place to permit an effective response and forensic investigation if an event or incident occurs.

---

[72] McKemmish, D. Rodney; *Computer and Intrusion Forensics*, Artech House, USA, 2003

Section 5: Incident Response

DATA ACQUISITION
All information and data required should be transferred into a controlled location; this includes all types of electronic media such as fixed disk drives and removable media. Each device must be checked to ensure that it is write-protected. This may be achieved by using a device known as a write-blocker. It is also possible to get data and information from witnesses or related parties by recorded statements. By volatile data, investigators can determine what is currently happening on a system. This kind of data includes open ports, open files, active processes, user logons and other data present in RAM. This information is lost when the computer is shut down.

IMAGING
Imaging is a process that allows one to obtain a bit-for-bit copy of data to avoid damage of original data or information when multiple analyses may be performed. The imaging process is made to obtain residual data, such as deleted files, fragments of deleted files and other information present, from the disk for analysis. This is possible because imaging duplicates the disk surface, sector by sector. With appropriate tools, it is sometimes possible to recover destroyed information (erased even by reformatting) from the disk's surface.

EXTRACTION
This process consists of identification and selection of data from the imaged data set. This process should include standards of quality, integrity and reliability. The extraction process includes software used and media where an image was made. The extraction process could include different sources such as system logs, firewall logs, intrusion detection system (IDS) logs, audit trails and network management information.

INTERVIEWS
Interviews are used to obtain prior indicators or relationships, including telephone numbers, IP addresses and names of individuals, from extracted data.

INGESTION/NORMALIZATION
This process converts the information extracted to a format that can be understood by investigators. It includes conversion of hexadecimal or binary data into readable characters or a format suitable for data analysis tools. It is possible to create relationships from data by extrapolation, using techniques such as fusion, correlation, graphing, mapping or time lining, which could be used in the construction of the investigation's hypothesis.

REPORTING
The information obtained from digital forensics has limited value when it is not collected and reported in the proper way. A report must state why the system was reviewed, how the computer data were reviewed and what conclusions were made from this analysis. The report should achieve the following goals:[73]
- Accurately describe the details of an incident.
- Be understandable to decision makers.
- Be able to withstand a barrage of legal scrutiny.
- Be unambiguous and not open to misinterpretation.
- Be easily referenced.
- Contain all information required to explain conclusions reached.
- Offer valid conclusions, opinions or recommendations when needed.
- Be created in a timely manner.

The report should also identify the organization, sample reports and restrictions on circulation (if any) and include any reservations or qualifications that the assurance professional has with respect to the assignment.

---

[73] Mandia, Kevin; Matt Pepe, Chris Prosise, *Incident Response & Computer Forensics, 2nd Edition*, McGraw Hill/Osborne, USA, 2003

NETWORK TRAFFIC ANALYSIS

Network traffic analysis identifies patterns in network communications. Traffic analysis does not need to have the actual content of the communication but analyzes where traffic is taking place, when and for how long communications occur, and the size of information transferred.

Traffic analysis can be used proactively to identify potential anomalies in communications or during incident response to develop footprints that identify different attacks or the activities of different individuals.

LOG FILE ANALYSIS

Many types of tools have been developed to help reduce the amount of information contained in audit records and to delineate useful information from the raw data. On most systems, audit trail software can create large files, which can be extremely difficult to analyze manually. The use of automated tools is likely to be the difference between unused audit trail data and an effective review. Some of the types of tools include:

- **Audit reduction tools**—These are preprocessors designed to reduce the volume of audit records to facilitate manual review. Before a security review, these tools can remove many audit records known to have little security significance. (This alone may cut in half the number of records in the audit trail.) These tools generally remove records generated by specified classes of events; for example, records generated by nightly backups might be removed.
- **Trend/variance-detection tools**—These look for anomalies in user or system behavior. It is possible to construct more sophisticated processors that monitor usage trends and detect major variations. For example, if a user typically logs in at 09.00, but appears at 04.30 one morning, this may indicate a security problem that may need to be investigated.
- **Attack-signature-detection tools**—These look for an attack signature, which is a specific sequence of events indicative of an unauthorized access attempt. A simple example would be repeated failed logon attempts.

DIGITAL FORENSIC TOOLS

Forensic tools can be sorted into four categories:

- **Computer**—Examine nonvolatile digital media. Due to the number of tools on the market, specific tools will not be discussed. Vendors base their tools on different platforms (i.e., Windows, Linux, etc.). Most are propriety; however, open source options do exist. Similarly, some are restricted to law enforcement and/or government agencies. Ultimately, business requirements will determine selection.
- **Memory**—Used to acquire and analyze volatile memory.
- **Mobile device**—Consists of both software and hardware components. Due to the wide number of devices, manufacturers and intended scope, specific tools will not be discussed. Cables perform similar to write blockers for computer forensics.
- **Network**—Monitoring and analysis of network traffic. Options range from command-line tools previously mentioned to high-end deep packet inspection appliances.

Additionally, there are various support applications that can be used. One example is VMware®—virtualization software that allows users to run multiple instances of operating systems on a physical PC or server.

TIME LINES

Time lines are chronological graphs where events related to an incident can be mapped to look for relationships in complex cases. Time lines can provide simplified visualization for presentation to management and other nontechnical audiences.

ANTI-FORENSICS

Programmers develop anti-forensics tools to make it difficult or impossible for investigators to retrieve information during an investigation. There are numerous ways people can hide information.

Anti-forensics tactics, techniques and procedures (TTPs) include:
- Securely deleting data
- Overwriting metadata
- Preventing data creation
- Encrypting data
- Encrypting network protocols
- Hiding data in slack space or other unallocated locations
- Hiding data or a file within another file (steganography)

TOPIC 5—DISASTER RECOVERY AND BUSINESS CONTINUITY PLANS

Disasters are disruptions that cause critical information resources to be inoperative for a period of time, adversely impacting organizational operations. The disruption could be a few minutes to several months, depending on the extent of damage to the information resource. Most important, disasters require recovery efforts to restore operational status.

A disaster may be caused by natural calamities, such as earthquakes, floods, tornadoes and fire, or a disaster may be caused by events precipitated by humans such as terrorist attacks, hacker attacks, viruses or human error. Many disruptions start as minor incidents. Normally, if the organization has a help desk or service desk, it would act as the early warning system to recognize the first signs of an upcoming disruption. Often, such disruptions (e.g., gradually deteriorating database performance) go undetected. Until these "creeping disasters" strike (the database halts), they cause only infrequent user complaints.

A cybersecurity-related disaster may occur when a disruption in service is caused by system malfunctions, accidental file deletions, untested application releases, loss of backup, network DoS attacks, intrusions or viruses. These events may require action to recover operational status in order to resume service. Such actions may necessitate restoration of hardware, software or data files.

BUSINESS CONTINUITY AND DISASTER RECOVERY

The purpose of business continuity planning (BCP)/disaster recovery planning (DRP) is to enable a business to continue offering critical services in the event of a disruption and to survive a disastrous interruption to activities. Rigorous planning and commitment of resources are necessary to adequately plan for such an event.

BCP takes into consideration:
- Critical operations necessary to the survival of the organization
- The human/material resources supporting these critical operations
- Predisaster readiness covering incident response management to address all relevant incidents affecting business processes
- Evacuation procedures
- Procedures for declaring a disaster (escalation procedures)
- Circumstances under which a disaster should be declared (Note: All interruptions are not disasters, but a small incident not addressed in a timely or proper manner may lead to a disaster. For example, a virus attack not recognized and contained in time may bring down the entire IT facility.)
- The clear identification of the responsibilities in the plan
- The clear identification of the persons responsible for each function in the plan
- The clear identification of contract information
- The step-by-step explanation of the recovery process
- The clear identification of the various resources required for recovery and continued operation of the organization

BCP is primarily the responsibility of senior management, because they are entrusted with safeguarding the assets and the viability of the organization, as defined in the BCP/DRP policy. The BCP is generally followed by the business and supporting units, to provide a reduced but sufficient level of functionality in the business operations immediately after encountering an interruption, while recovery is taking place.

Depending on the complexity of the organization, there could be one or more plans to address the various aspects of BCP and DRP. These plans do not necessarily have to be integrated into one single plan. However, each has to be consistent with other plans to have a viable BCP strategy.

Even if similar processes of the same organization are handled at a different geographic location, the BCP and DRP solutions may be different for different scenarios. Solutions may be different due to contractual requirements (e.g., the same organization is processing an online transaction for one client and the back office is processing for another client). A BCP solution for the online service will be significantly different than one for the back office processing.

Section 5: Incident Response

BUSINESS IMPACT ANALYSIS

The first step in preparing a new BCP is to identify the business processes of strategic importance—those key processes that are responsible for both the permanent growth of the business and for the fulfillment of the business goals. Ideally, the BCP/DRP should be supported by a formal executive policy that states the organization's overall target for recovery and empowers those people involved in developing, testing and maintaining the plans.

Based on the key processes, a business impact analysis (BIA) process should begin to determine time frames, priorities, resources and interdependencies that support the key processes. Business risk is directly proportional to the impact on the organization and the probability of occurrence of the perceived threat. Thus, the result of the BIA should be the identification of the following:
- The human resources, data, infrastructure elements and other resources (including those provided by third parties) that support the key processes
- A list of potential vulnerabilities—the dangers or threats to the organization
- The estimated probability of the occurrence of these threats
- The efficiency and effectiveness of existing risk mitigation controls (risk countermeasures)

Information is collected for the BIA from different parts of the organization which own key processes/applications. To evaluate the impact of downtime for a particular process/application, the impact bands are developed (i.e., high, medium, low) and, for each process, the impact is estimated in time (hours, days, weeks). The same approach is used when estimating the impact of data loss. If necessary, the financial impact may be estimated using the same techniques, assigning the financial value to the particular impact band. In addition, data for the BIA may be collected on the time frames needed to supply vital resources—how long the organization may run if a supply is broken or when the replacement has arrived.

The BIA should answer three important questions:
1. What are the different business processes?
2. What are the critical information resources related to an organization's critical business processes?
3. What is the critical recovery time period for information resources in which business processing must be resumed before significant or unacceptable losses are suffered?

The core source of data used in BCP is the BIA, which identifies the critical time lines for services and products associated with value creation and business risk tolerance. The BIA also establishes the recovery point objective (RPO) and recovery time objective (RTO) for a process, which respectively defines how much data can be lost in recovery and how quickly recovery must be accomplished. Because the BCP is based on the BIA, the risk practitioner must review the process used to determine the BIA to validate that it is accurate and considers all of the risk factors. In organizations that have adopted a comprehensive business continuity management system (BCMS), the risk practitioner may benefit from having direct access to BCPs for review and reference in the context of risk management.[74]

SUPPLY CHAIN CONSIDERATIONS

NIST defines the information and communications technology (ICT) supply chain as "a complex, globally distributed, and interconnected ecosystem that is long, has geographically diverse routes, and consists of multiple tiers of outsourcing." This environment is interdependent on public and private entities for development, integration and delivery of ICT products and services.[75]

The complexity of supply chains and impact requires persistent awareness of risk and consideration. The most significant factors contributing to the fragility of supply chains are economic, environmental, geopolitical and technological.[76]

---

[74] ISACA, *CRISC Review Manual 6th Edition*, USA, 2015
[75] Boyens, Jon; Celia Paulsen; Rama Moorthy; Nadya Bartol; *NIST SP 800-161, 2nd draft: Supply Chain Risk Management Practices for Federal Information Systems and Organization*, NIST, USA, 2014
[76] Rodrigue, Jean-Paul; "Risk in Global Supply Chains," https://people.hofstra.edu/geotrans/eng/ch9en/conc9en/supply_chain_risks.html

Whether it is the rapid adoption of open-source software, tampering of physical hardware or natural disasters taking down data centers, supply chains require risk management. An example of this was described in an article in Forbes: Flooding in Thailand created significant shortages in the hard disk drive market, which cost well-known electronics manufacturers millions of dollars in losses.[77]

Products or services manufactured anywhere may contain vulnerabilities that can present opportunities for ICT supply chain-related compromises. It is especially important to consider supply chain risk from system development, to include research and development (R&D) through useful life and ultimately retirement/disposal of products.

IS BUSINESS CONTINUITY PLANNING

In the case of IS BCP, the approach is the same as in BCP with the exception being that the continuity of IS processing is threatened. IS processing is of strategic importance—it is a critical component because most key business processes depend on the availability of key systems, infrastructure components and data.

The IS BCP should be aligned with the strategy of the organization. The criticality of the various application systems deployed in the organization depends on the nature of the business as well as the value of each application to the business.

The value of each application to the business is directly proportional to the role of the information system in supporting the strategy of the organization. The components of the information system (including the technology infrastructure components) are then matched to the applications (e.g., the value of a computer or a network is determined by the importance of the application system that uses it).

Therefore, the information system BCP/DRP is a major component of an organization's overall business continuity and disaster recovery strategy. If the IS plan is a separate plan, it must be consistent with and support the corporate BCP. See **figure 5.5**.

Figure 5.5—Business Continuity Planning

Business Impact Analysis provides basis for
↓
Business Continuity Planning, which determines
↓
Recovery Time Objectives
Recovery Point Objectives
Maximum Tolerable Outages
Service Delivery Objectives

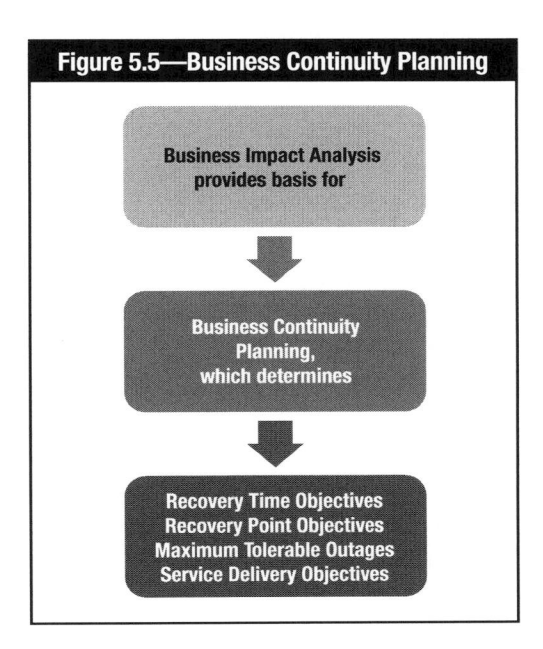

[77] Culp, Steve, "Supply Chain Risk a Hidden Liability for Many Companies," *Forbes*, 8 October 2012, *www.forbes.com/sites/steveculp/2012/10/08/supply-chain-risk-a-hidden-liability-for-many-companies*

RECOVERY CONCEPTS

Data recovery is the process of restoring data that has been lost, accidentally deleted, corrupted or made inaccessible for any reason. Recovery processes vary depending on the type and amount of data lost, the backup method employed and the backup media. An organization's DRP must provide the strategy for how data will be recovered and assign recovery responsibilities.

BACKUP PROCEDURES

Backup procedures are used to copy files to a second medium such as a disk, tape or the cloud. Backup files should be kept at an offsite location. Backups are usually automated using operating system commands or backup utility programs. Most backup programs compress the data so that the backups require fewer media.

There are three types of data backups: full, incremental and differential. Full backups provide a complete copy of every selected file on the system, regardless of whether it was backed up recently. This is the slowest backup method but the fastest method for restoring data. Incremental backups copy all files that have changed since the last backup was made, regardless of whether the last backup was a full or incremental backup. This is the fastest backup method but the slowest method for restoring data. Differential backups copy only the files that have changed since the last full backup. The file grows until the next full backup is performed.

SECTION 5—KNOWLEDGE CHECK

1. Arrange the steps of the incident response process into the correct order.
   A. Mitigation and recovery
   B. Investigation
   C. Postincident analysis
   D. Preparation
   E. Detection and analysis

2. Which element of an incident response plan involves obtaining and preserving evidence?
   A. Preparation
   B. Identification
   C. Containment
   D. Eradication

3. Select three. The chain of custody contains information regarding:
   A. disaster recovery objectives, resources and personnel.
   B. who had access to the evidence, in chronological order.
   C. labor, union and privacy regulations.
   D. proof that the analysis is based on copies identical to the original evidence.
   E. the procedures followed in working with the evidence.

4. NIST defines a(n) _____ as a "violation or imminent threat of violation of computer security policies, acceptable use policies, or standard security practices."
   A. Disaster
   B. Event
   C. Threat
   D. Incident

5. Select all that apply. A business impact analysis (BIA) should identify:
   A. the circumstances under which a disaster should be declared.
   B. the estimated probability of the identified threats actually occurring.
   C. the efficiency and effectiveness of existing risk mitigation controls.
   D. a list of potential vulnerabilities, dangers and/or threats.
   E. which types of data backups (full, incremental and differential) will be used.

See answers in Appendix C.

Page intentionally left blank

Section 6: Security Implications and Adoption of Evolving Technology

Topics covered in this section include:
1. Current threat landscape
2. Advanced persistent threats (APTs)
3. Mobile technology—vulnerabilities, threats and risk
4. Consumerization of IT and mobile devices
5. Cloud and digital collaboration

Page intentionally left blank

TOPIC 1—CURRENT THREAT LANDSCAPE

A threat landscape, also referred to as a threat environment, is a collection of threats. The cybersecurity threat landscape is constantly changing and evolving as new technologies are developed and cyberattacks and tools become more sophisticated. A threat landscape developed by the European Union Agency for Network and Information Security (ENISA) is shown in **figure 6.1**. Corporations are becoming increasingly dependent on digital technologies that can be susceptible to cybersecurity risk. Cloud computing, social media and mobile computing are changing how organizations use and share information. They provide increased levels of access and connectivity, which create larger openings for cybercrime.

Figure 6.1—ENISA Threat Landscape 2015

Top Threats 2015	Assessed Trends 2015	Change in ranking
1. Malware	Increasing	Same
2. Web based attacks	Increasing	Same
3. Web application attacks	Increasing	Same
4. Botnets	Declining	Same
5. Denial of service	Increasing	Same
6. Physical damage/theft/loss	Stable	Going up
7. Insider threat (malicious, accidental)	Increasing	Going up
8. Phishing	Stable	Going down
9. Spam	Declining	Going down
10. Exploit kits	Increasing	Going down
11. Data breaches	Stable	Going down
12. Identify theft	Stable	Going up
13. Information leakage	Increasing	Going down
14. Ransomware	Increasing	Going up
15. Cyber espionage	Increasing	Going down

Legend: Trends: ⬆ Increasing, ➡ Stable, ⬇ Declining
Ranking: ↑ Going up, → Same, ↓ Going down

Source: ENISA, *ENISA Threat Landscape 2015*, Greece, 2016

Cybercriminals are usually motivated by one or more of the following:
- Financial gains
- Intellectual property (espionage)
- Politics (hacktivism)

Recent trends in the cyberthreat landscape include:
- Threat agents are more sophisticated in their attacks and use of tools.
- Attack patterns are now being applied to mobile devices. This is of particular concern for mobile and other small digital devices that are interconnected and often have poor security controls.
- Multiple nation states have the capabilities to infiltrate government and private targets (cyberwarfare).
- Cloud computing results in large concentrations of data within a small number of facilities, which are likely targets for attackers.
- Social networks have become a primary channel for communication, knowledge collection, marketing and dissemination of information. Attackers can misuse social networks to gain personal data and promulgate misinformation.
- Big data refers to large collections of structured and unstructured data and the usage of large infrastructure, applications, web services and devices. The popularity of big data as an asset allows for the potential for big data breaches.

TOPIC 2—ADVANCED PERSISTENT THREATS

Advanced persistent threats (APTs) are relatively new phenomena for many organizations. Although the motives behind them are not entirely new, the degree of planning, resources employed and techniques used in APT attacks are unprecedented. These threats demand a degree of vigilance and a set of countermeasures that are above and beyond those routinely used to counter everyday security threats from computer hackers, viruses or spammers.[78] **Figure 6.2** shows the evolution of the threat landscape.

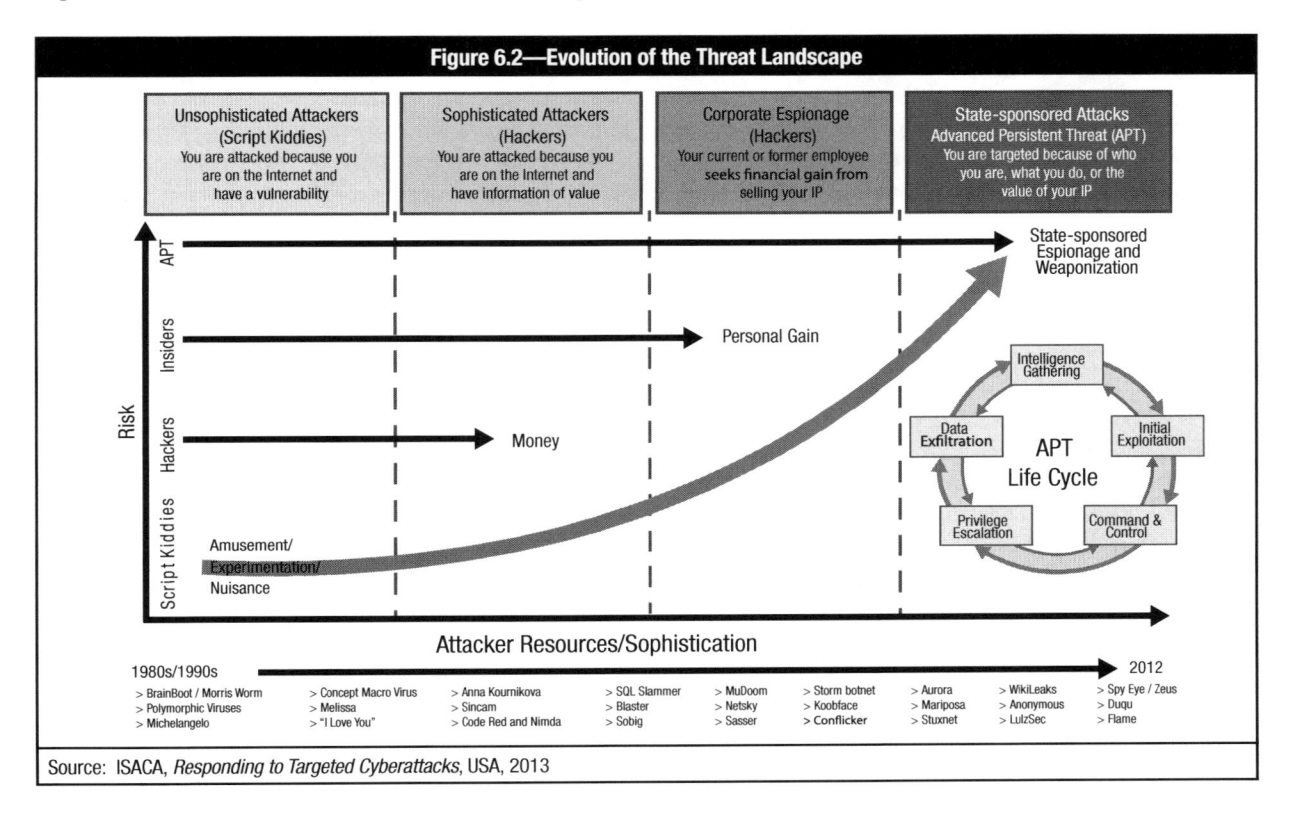

Figure 6.2—Evolution of the Threat Landscape

Source: ISACA, *Responding to Targeted Cyberattacks*, USA, 2013

DEFINING APTS

It should be noted that not everyone agrees on precisely what constitutes an APT. Many experts regard it as nothing new. Some see it as simply the latest evolution in attack techniques that have been developing over many years. Others claim the term is misleading, pointing out that many attacks classed as APTs are not especially clever or novel. A few define it in their own terms; for example, as an attack that is professionally managed; as one that follows a particular *modus operandi*; as one launched by a foreign intelligence service; or as one that targets and relentlessly pursues a specific enterprise.

In fact, all of these descriptions are true. The defining characteristics of an APT are very simple: An **APT** is a targeted threat that is composed of various complex attack vectors and can remain undetected for an extended period of time. It is a specifically targeted and sophisticated attack that keeps coming after the victim. Unlike many other types of criminal acts, it is not easily deflected by a determined, defensive response. An example of an APT is spear phishing, where social engineering techniques are used to masquerade as a trusted party to obtain important information such as passwords from the victim.

---

[78] ISACA, *Advanced Persistent Threats: How to Manage the Risk to Your Business*, USA, 2013

APT CHARACTERISTICS

Attacks of this kind are quite different from the ones that enterprises might have experienced in the past. Most organizations have at some point encountered one or more opportunistic attacks from small-time criminals, hackers or other mischief makers. Most APT attacks originate from more sinister sources. They are often the work of professional teams employed by organized crime groups, determined activists or governments. This means they are likely to be well-planned, sophisticated, well-resourced and potentially more damaging.

APT attacks vary significantly in their approach; however, they share the following characteristics:
- **Well-researched**—APT agents thoroughly research their targets, plan their use of resources and anticipate countermeasures.
- **Sophisticated**—APT attacks are often designed to exploit multiple vulnerabilities in a single attack. They employ an extensive framework of attack modules designed for executing automated tasks and targeting multiple platforms.
- **Stealthy**—APT attacks often go undetected for months and sometimes years. They are unannounced and disguise themselves using obfuscation techniques or hide in out-of-reach places.
- **Persistent**—APT attacks are long-term projects with a focus on reconnaissance. If one attack is successfully blocked, the perpetrators respond with new attacks. Plus, they are always looking for methods or information to launch future attacks.

APT TARGETS

APTs target companies of all sizes across all sectors of industry and all geographic regions that contain high-value assets. Staff of all levels of seniority, ranging from administrative assistants to chief executives, can be selected as a target for a spear phishing attack. Small companies and contractors might be penetrated because they are a supplier of services to a targeted victim. Individuals might be selected if they are perceived to be a potential stepping stone to help gain access to the ultimate target.

No industry with valuable secrets or other sources of commercial advantage that can be copied or undermined through espionage is safe from an APT attack. No enterprise that controls money transfers, processes credit card data or stores personally identifiable data on individuals can be sheltered from criminal attacks. Likewise, no industry that supplies or supports critical national infrastructure is immune from an intrusion by cyberwarriors.

APT attacks often encompass third-party organizations delivering services to targeted enterprises. Third-party suppliers can be perceived by an attacker as the weakest link of large companies and government departments because they are generally less protected. No matter how effective a company's external perimeter security might be, it can be of limited value unless it is extended across its supply chain.

Figure 6.3 lists the primary actors behind APT threats. It sets out their overall goals as well as the potential business impact of their attacks.

Figure 6.3—APT Types and Impacts		
Threat	**What They Seek**	**Business Impact**
Intelligence agencies	Political, defense or commercial trade secrets	Loss of trade secrets or commercial, competitive advantage
Criminal groups	Money transfers, extortion opportunities, personal identity information or any secrets for potential onward sale	Financial loss, large-scale customer data breach or loss of trade secrets
Terrorist groups	Production of widespread terror through death, destruction and disruption	Loss of production and services, stock market irregularities, and potential risk to human life
Activist groups	Confidential information or disruption of services	Major data breach or loss of service
Armed forces	Intelligence or positioning to support future attacks on critical national infrastructure	Serious damage to facilities in the event of a military conflict

STAGES OF AN APT ATTACK

Even though no two APT attacks are exactly alike, they often follow a similar life cycle, shown in **figure 6.4.** They start with intelligence gathering, which includes selecting and researching their target, planning the attack and collecting and analyzing data from an initial penetration. The attacker then establishes command and control, collecting targeted information. That information is then exfiltrated to the attacker's location to be disseminated or exploited.

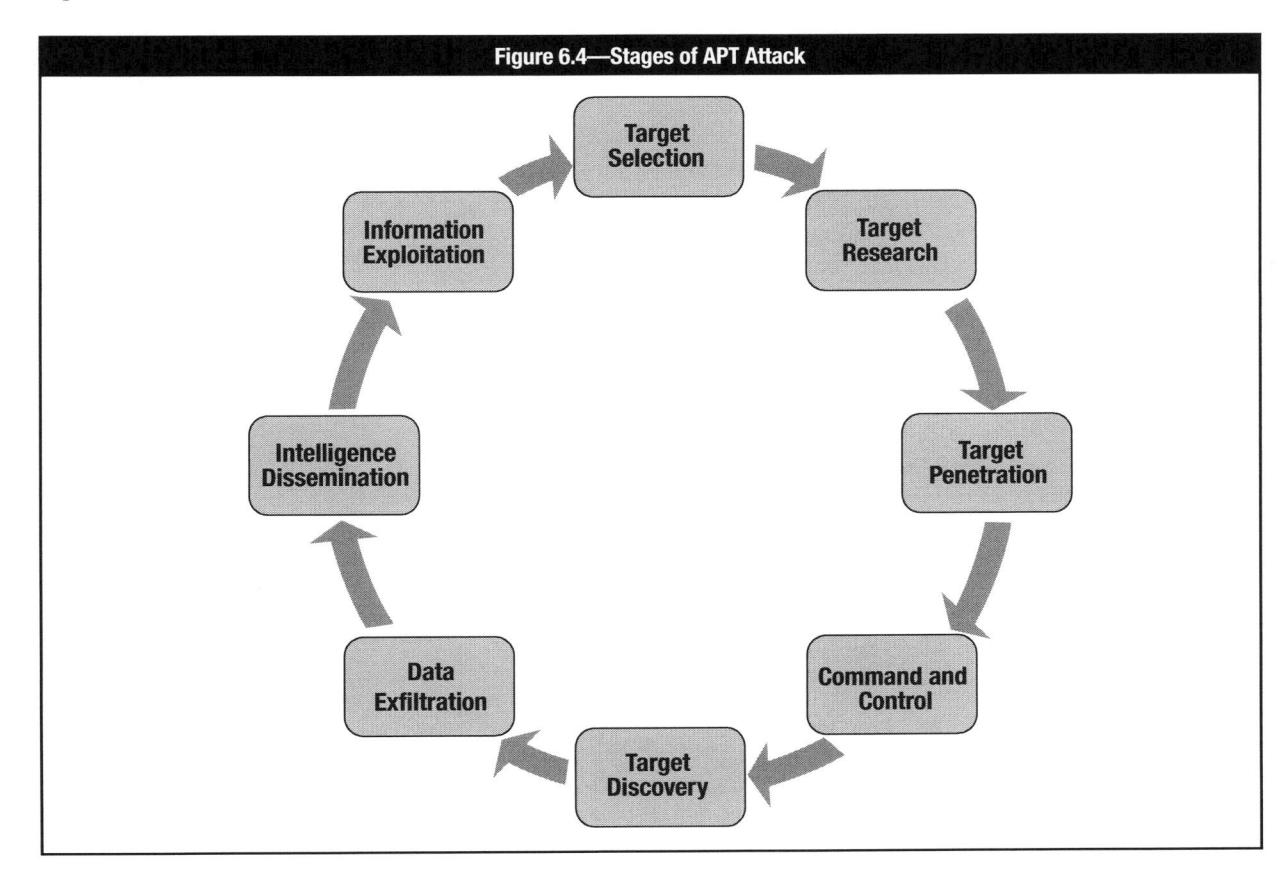

Figure 6.4—Stages of APT Attack

- Target Selection
- Target Research
- Target Penetration
- Command and Control
- Target Discovery
- Data Exfiltration
- Intelligence Dissemination
- Information Exploitation

Page intentionally left blank

TOPIC 3—MOBILE TECHNOLOGY—VULNERABILITIES, THREATS AND RISK

Security for mobile technology is a function of the risk associated with its use. Despite positive and negative impacts, security teams must deal with the risk common to all mobile devices and applications.

MOBILE APPLICATIONS

Similar to the list discussed in section 4, topic 6, OWASP provides a Mobile Top Ten that defines the top 10 critical mobile security flaws that should be addressed by cybersecurity teams when dealing with mobile devices and applications (**figure 6.5**)

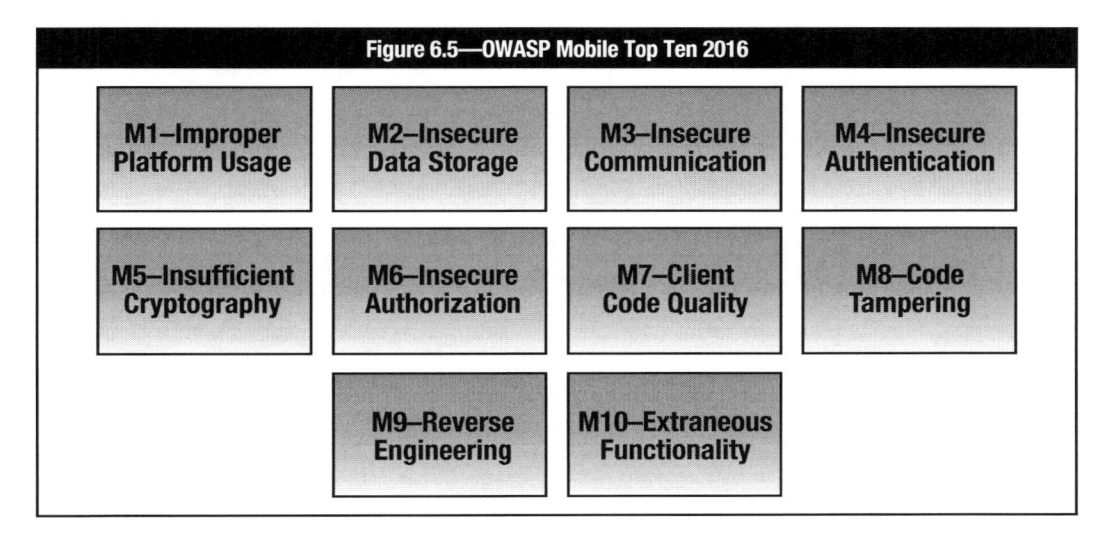

Figure 6.5—OWASP Mobile Top Ten 2016

- M1—Improper Platform Usage
- M2—Insecure Data Storage
- M3—Insecure Communication
- M4—Insecure Authentication
- M5—Insufficient Cryptography
- M6—Insecure Authorization
- M7—Client Code Quality
- M8—Code Tampering
- M9—Reverse Engineering
- M10—Extraneous Functionality

PHYSICAL RISK

Mobile devices tend to be small, by definition. They are easily lost (or stolen), particularly in public areas. This increases the general physical risk, given that advanced smartphones are often seen as interesting targets for pickpockets. As users increasingly rely on their mobile devices, loss or theft is more likely to create disruptive conditions and may leave employees unable to work for prolonged periods of time.

The security consequences of device loss are more serious. For example, unprotected and transient data, such as lists of calls, texts or calendar items, may be compromised, allowing attackers to harvest large amounts of data. With criminal intent, perpetrators may be able to recover deleted data and a history of the use of the mobile device.

An additional significant risk is identity theft, which may occur as a result of obtaining and analyzing a stolen or lost mobile device. Many mainstream operating systems (OSs) for smart devices mandate the link to a user account with the provider, thus greatly increasing the risk of losing one's digital identity with the actual device.

The link between device and account is sometimes subject to even greater risk when value-added services are offered as an add-on to the existing user account. Some OSs offer a "secure" repository for enriched user data ranging from personal information to automated credit card storage and payment functionality. The risk of entrusting such sensitive data to a mobile device ("all in one place") should not be neglected.

Section 6: Security Implications and Adoption of Evolving Technology

From a security management perspective, several attempts have been undertaken to prevent, or at least mitigate, the threat of device loss or theft:
- Cell-based tracking and locating the device
- Remote shutdown/wipe capabilities
- Remote SIM card lock capabilities

While these facilities do provide a degree of security, they still leave a window of exposure to attackers exploring the device, possibly using analytical tools that will circumvent the standard OS features. This threat is particularly significant because enforcing strong passwords and encryption on mobile devices may be restricted due to OS limitations.

ORGANIZATIONAL RISK

As with many other technologies, mobile devices have rapidly pervaded enterprises at all levels. They are now available to most users, either through corporate provisioning or bring your own device (BYOD) programs. In terms of data, information and knowledge that exist across the enterprise, many users have privileged access that is often replicated on their mobile devices.

Whereas corporate PC environments have been the target of hardening and protective measures for many years, mobile devices and their comparatively weak security mechanisms are more difficult to manage and control. As an example, C-suite and senior managers will often be heavy mobile users, and any successful compromise of their devices could certainly cause major damage.

Another important organizational risk arises from the growing complexity and diversity of common mobile devices. Whereas early cell phones required no more than the most basic knowledge about how to use a keyboard, smartphones offer everything from simple telephony to highly complex applications. Even for experienced users, this level of complexity may be challenging, and many smartphones are thought to be conducive to human error and user-based security issues. Examples such as inadvertent data roaming or involuntary GPS tagging show how many users simply do not understand the extended features of their devices.

At the same time, the rapid succession of new generations of hardware requires constant adaptation on the part of users and enterprises. The comparatively long systems management cycles found in larger enterprises may cause difficulties when facing the usual turnaround time of approximately two years for new mobile devices. Likewise, the life span of mobile OSs and applications is becoming much shorter.

The resulting risk to users is aggravated by the fact that few enterprises offer formal or informal training for mobile device use. Users are often left on their own when it comes to adopting and using new technology and new services.

TECHNICAL RISK

Activity Monitoring and Data Retrieval

In general, mobile devices use service-based OSs with the ability to run multiple services in the background. While early variants of these OSs were fairly transparent and controllable in terms of activity, more recent versions show a tendency to "simplify" the user interface by restricting the user's ability to change low-level settings. However, monitoring and influencing activity is a core functionality of spyware and malware, as is covert data retrieval. Data can be intercepted in real time as they are being generated on the device. Examples include sending each email sent on the device to a hidden third-party address, letting an attacker listen in on phone calls or simply opening microphone recording. Stored data such as a contact list or saved email messages can also be retrieved. **Figure 6.6** shows an overview of targets and the corresponding risk.

Section 6: Security Implications and Adoption of Evolving Technology

Figure 6.6—Activity Monitoring and Data Retrieval Risk[79]

Target	Risk
Messaging	Generic attacks on SMS text, MMS-enriched transmission of text and contents
	Retrieval of online and offline email contents
	Insertion of service commands by SMS cell broadcast texts
	Arbitrary code execution via SMS/MMS
	ML-enabled SMS text or email
	Redirect or phishing attacks by HTML-enabled SMS text or email
Audio	Covert call initiation, call recording
	Open microphone recording
Pictures/Video	Retrieval of still pictures and videos, for instance, by piggybacking the usual "share" functionality in most mobile apps
	Covert picture or video taking and sharing, including traceless wiping of such material
Geolocation	Monitoring and retrieval of GPS positioning data, including date and time stamps
Static data	Contact list, calendar, tasks, notes retrieval
History	Monitoring and retrieval of all history files in the device or on SIM card (calls, SMS, browsing, input, stored passwords, etc.)
Storage	Generic attacks on device storage (hard disk or solid state disk [SSD]) and data replicated there

In practice, the risk has already materialized for most common device platforms and OSs.

In combination with attacks on connectivity, the risk of activity monitoring/influencing and covert data retrieval is significant.

UNAUTHORIZED NETWORK CONNECTIVITY

Most spyware or malware—once placed on a mobile device—requires one or more channels for communicating with the attacker. While "sleepy" malware may have a period of latency and remain dormant for weeks or months, data and information harvested will eventually need to be transmitted from the mobile device to another destination.

Similarly, the command and control functionality often found in malware requires a direct link between the mobile device and the attacker, particularly when commands and actions are to be executed and monitored in real time (e.g., eavesdropping or covert picture taking). **Figure 6.7** shows the most common vectors for unauthorized network connectivity and the typical risk that comes with them.

Figure 6.7—Unauthorized Connectivity Risk[80]

Vector	Risk
Email	Simple to complex data transmission (including large files)
SMS	Simple data transmission, limited command and control (service command) facility
HTTP get/post	Generic attack vector for browser-based connectivity, command and control
TCP/UDP socket	Lower-level attack vector for simple to complex data transmission
DNS exfiltration	Lower-level attack vector for simple to complex data transmission, slow but difficult to detect
Bluetooth	Simple to complex data transmission, profile-based command and control facility, generic attack vector for close proximity
WLAN/WiMAX	Generic attack vector for full command and control of target, equivalent to wired network

---

[79] ISACA, *Securing Mobile Devices*, USA, 2012
[80] *Ibid*

Section 6: Security Implications and Adoption of Evolving Technology

These vectors of connectivity may be used in combination, for example, when browser-based command and control functionality is used via Bluetooth in a close-proximity scenario. An important point to note is the relative anonymity of wireless connectivity vectors, particularly Bluetooth and WLAN/WiMAX. The risk of ad hoc attacks on mobile devices is significantly higher when anonymous connectivity is provided by third parties, for example, in airport lounges or coffee shops.

WEB VIEW/USER INTERFACE (UI) IMPERSONATION

While most mobile devices support all relevant browser protocols, the presentation to the user is modified by the mobile service provider. This is mainly done to optimize viewing on small screens. However, web pages viewed on a typical (smaller) device often show "translated" content, including modifications to the underlying code.

In UI impersonation, malicious apps present a UI that impersonates the native device or that of a legitimate app. When the victim supplies authentication credentials, these are transmitted to the attacker. This is conducive to impersonation attacks that are similar to generic phishing.

Typical web view applications allow attacks on the proxy level (phishing credentials while proxying to a legitimate website) and on the presentation level (fake website presented through mobile web view). This type of risk is prevalent in banking applications where several cases of malware have been documented. Given the attractiveness of payment data and user credentials, web view and impersonation risk is likely to increase in the future.

SENSITIVE DATA LEAKAGE

With the emergence of new work patterns and the need for decentralized data availability, mobile devices often store large amounts of sensitive data and information. As an example, confidential presentations and spreadsheets are often displayed directly from a smart mobile device rather than using a laptop computer.

The amount of storage space found on many devices is growing and, on average, almost any device will soon be capable of storing several gigabytes of data. This greatly increases the risk of data leakage, particularly when mobile devices store replicated information from organizational networks. This is often the case when standard email and calendar applications automatically replicate emails with attachments, or mobile OSs offer the convenience of replicating selected folders between mobile device and desktop device. **Figure 6.8** shows the information targeted and possible risk.

Figure 6.8—Sensitive Data Leakage Risk[81]	
Type of Information	**Risk**
Identity	International Mobile Equipment Identity (IMEI), manufacturer device ID, customized user information
	Hardware/firmware and software release stats, also disclosing known weaknesses or potential zero-day exploits
Credentials	User names and passwords, keystrokes
	Authorization tokens, certificates (S/MIME, PGP, etc.)
Location Files	GPS coordinates, movement tracking, location/behavioral inference
	All files stored at operating system/file system level

Sensitive data leakage can be inadvertent or can occur through side channel attacks. Even a legitimate application may have flaws in the usage of the device. As a result, information and authentication credentials may be exposed to third parties. Depending on the nature of the information leaked, additional risk may arise.

Mobile devices provide a fairly detailed picture of what their users do, where they are and what their preferences are. Side channel attacks over prolonged periods of time allow the building of a detailed user profile in terms of movements, behavior and private/business habits. Users who may be considered at risk may require additional physical protection.

[81] *Ibid*

Sensitive data leakage allowing the prediction of users' behavior patterns and activities is becoming more significant as many users prefer to set their devices to "always on" mode to benefit from legitimate services such as navigation or local points of interest.

UNSAFE SENSITIVE DATA STORAGE

While most mobile OSs offer protective facilities such as storage encryption, many applications store sensitive data such as credentials or tokens as plaintext. Furthermore, data stored by the user is often replicated without encryption, and many standardized files such as Microsoft Office® presentations and spreadsheets are stored unencrypted for quick access and convenience.

Another risk associated with unsafe storage of sensitive data is the use of public cloud services for storage purposes. Many mobile device providers have introduced cloud services that offer a convenient way of storing, sharing and managing data in a public cloud. However, these services target the private consumer, and the security functionality would not normally stand up to organizational (corporate) requirements.

This risk has another dimension: when data and information are stored or replicated in public clouds, terms and conditions generally rule out any form of responsibility or liability, requiring the user to make individual security arrangements. In an organizational context, these limitations may increase the risk of sensitive data storage, particularly in a BYOD scenario.

UNSAFE SENSITIVE DATA TRANSMISSION

Mobile devices predominantly rely on wireless data transmission, except for the few cases when they are physically connected to a laptop or desktop computer. As outlined previously, these transmissions create a risk of unauthorized network connectivity, particularly when using a wireless local area network (WLAN)/worldwide interoperability for microwave access (WiMAX) or Bluetooth at close proximity. As a new transmission protocol, near-field communication (NFC) increases the risk at very short range, for example, when transmitting payment data over a distance of several inches.

Even if data at rest are protected by encryption and other means, transmission is not always encrypted. Mobile users are likely to use unsecured public networks frequently, and the large number of known attacks on WLAN and Bluetooth are a significant risk.

Automatic network recognition, a common feature in mobile OSs, may link to WLANs available in the vicinity, memorizing Service Set Identifiers (SSIDs) and channels. For many major providers of public WLANs, these SSIDs are identical across the world. This is intentional and convenient; however, the risk of an evil twin attack increases with the use of generic names that the mobile device will normally accept without verification.

While many enterprises have implemented virtual private network (VPN) solutions for their users, these may not be workable on mobile devices that are used both for business and personal transactions. Given the relative complexity of configuring and activating VPN on mobile devices, users may deactivate protected data transmission to access another service that does not support VPN. Even for split-tunnel VPN installations—offering a VPN to the enterprise while keeping the open link to the public network—the risk of an at-source attack is still high.

DRIVE-BY VULNERABILITIES

In contrast to desktop or laptop computers, mobile devices offer only rudimentary applications for office-based work. In many cases, device size restricts the display and edit capabilities. As a consequence, typical word processing, spreadsheet and presentation software on mobile devices tends to be optimized for opening and reading rather than editing information. Similarly, popular document formats such as Adobe® portable document format (PDF) are implemented, more or less, as a read-only solution designed for a cursory read rather than full-scale processing.

At the same time, it has become common practice to insert active content into documents and PDF files. These may be full hyperlinks or shortened links, or embedded documents and macros. This is known as an attack vector for malware and other exploits.

The restricted nature of mobile device applications leads to an increased risk of drive-by attacks because these apps may not recognize malformed links and omit the usual warnings that users could expect from the desktop versions of Microsoft Office or PDF applications.

In practice, these vulnerabilities create risk and a number of threats for end users, for example, the insertion of illegal material, inadvertent use of "premium" services via SMS/MMS or bypassing two-factor authentication mechanisms.

TOPIC 4—CONSUMERIZATION OF IT AND MOBILE DEVICES

CONSUMERIZATION OF IT

Mobile devices have had a profound impact on the way business is conducted and on behavior patterns in society. They have greatly increased productivity and flexibility in the workplace, to the extent that individuals are now in a position to work from anywhere at any given time. Likewise, the computing power of smart devices has enabled them to replace desktop PCs and laptops for many business applications.

Manufacturers and service providers alike have created both new devices and new business models such as mobile payments or subscription downloads using a pay-as-you-go model. Simultaneously, consumerization of devices has relegated enterprises, at least in some cases, to followers rather than opinion leaders in terms of which devices are used and how they are used.

The impact of using mobile devices falls into two broad categories:
- The hardware itself has been developed to a level at which computing power and storage are almost equivalent to PC hardware. In accordance with Moore's Law, a typical smartphone represents the equivalent of what used to be a midrange machine a decade ago.
- New mobile services have created new business models that are changing organizational structures and society as a whole.

Consumerization is not limited to devices. New, freely available applications and services provide better user experiences for things like note-taking, video conferencing, email and cloud storage than their respective corporate-approved counterparts. Instead of being provided with company-issued devices and software, employees are using their own solutions that better fit with their lifestyle, user needs and preferences.

BRING YOUR OWN DEVICE

General mobility and location-independent accessibility have enhanced business practices and have allowed enterprises to focus on core activities while reducing the amount of office space used. For employees, mobile devices have brought greater flexibility, for example, in bring your own device (BYOD) programs.

The idea of using privately owned mobile devices has quickly taken hold as a concept, and many enterprises are now facing a new obstacle: when centralized procurement and provisioning of mobile devices are slow or cumbersome, many users have developed the expectation of simply "plugging in" their own units to achieve productivity in a quick and pragmatic manner.

The obvious downside is the proliferation of devices with known (or unknown) security risk, and the formidable challenge of managing device security against several unknowns. However, as the workforce changes, there are clear signs that BYOD is becoming an important job motivation factor, because employees are no longer willing to accept technology restrictions.

While BYOD may be seen as an enabler, it has also brought a number of new risk areas and associated threats. These need to be balanced with the advantages of mobile device use, taking into account the security needs of the individual as well as the enterprise. Therefore, security management should address both the innovative potential and the risk and threats of flexible device use because it is unlikely that restrictions or bans on certain types of devices will be effective even in the medium term. Indeed, the fact that some enterprises have attempted a ban on certain devices has allowed the prohibited technology to gain a foothold within the corporate landscape—particularly if that technology is already widely accepted among private users. As a result, enterprises with a restrictive perspective on innovative devices will always be behind the threat curve and thus exposed to unnecessary risk. The pros and cons of BYOD are listed in **figure 6.9**.

Figure 6.9—Pros and Cons of BYOD

Pros	Cons
• Shifts costs to user	• IT loss of control
• Worker satisfaction	• Known or unknown security risk
• More frequent hardware upgrades	• Acceptable Use Policy is more difficult to implement
• Cutting-edge technology with the latest features and capabilities	• Unclear compliance and ownership of data

INTERNET OF THINGS (IOT)[82]

The Internet of Things (IoT) refers to physical objects that have embedded network and computing elements and communicate with other objects over a network. Definitions of IoT vary about the pathway of communication. Some definitions state that IoT devices communicate over the Internet; others state that IoT devices communicate via a network, which may or may not be the Internet.

This growing internetwork of "things" is comprised of physical objects with the capability to communicate in new ways—with each other, with their owners or operators, with their manufacturers or with others—to make people's lives easier and enterprises more efficient and competitive.

The IoT trend is transformative from a business standpoint. Business value and organizational competitiveness can be derived as enterprises capitalize on these new capabilities to gain more and better business value from devices that they purchase. Additionally, businesses can compete more effectively in the marketplace as they provide these features in products that they sell and incorporate them into service offerings that they provide.

However, with that additional value comes additional risk. Although specific risk depends on usage, some of the IoT usage risk areas that practitioners should consider are:
- Business risk:
  - Health and safety
  - Regulatory compliance
  - User privacy
  - Unexpected costs
- Operational risk:
  - Inappropriate access to functionality
  - Shadow usage
  - Performance
- Technical risk:
  - Device vulnerabilities
  - Device updates
  - Device management

---

[82] ISACA, *Internet of Things: Risk and Value Considerations*, USA, 2015

Figure 6.10 illustrates some points to consider regarding IoT.

Figure 6.10—IoT Dos and Don'ts

Dos	Don'ts
• Preprare a threat model. • Evaluate business value. • Holisticaly evalute and manage risk. • Balance risk and rewards. • Notify all stakeholders of anticipated usage. • Engage with business teams early. • Gather all stakeholders to ensure engagement and thorough planning. • Look for points of integration with existing security and operational protections. • Examine and document information that is collected and transmitted by devices to analyze possible privacy impacts. • Discuss with relevant stakeholders when, how and with whom that information will be shared and under what circumstances.	• Deploy quickly without consulting business or other stakeholders. • Disregard existing policy requirements, such as security and privacy. • Ignore regulatory mandates. • Assume vendors (hardware, software, middleware or any other) have thought through your particular usage or security requirements. • Disregard device-specific attacks or vulnerabilities. • Discount privacy considerations or "hide" data that are collected/transmitted from end users.

BIG DATA

Big data is both a marketing and a technical term referring to a valuable enterprise asset—information. Big data represents a trend in technology that is leading the way to a new approach in understanding the world and making business decisions. These decisions are made based on very large amounts of structured, unstructured and complex data (e.g., tweets, videos, commercial transactions) which have become difficult to process using basic database and warehouse management tools. Managing and processing the ever-increasing data set requires running specialized software on multiple servers. For some enterprises, big data is counted in hundreds of gigabytes; for others, it is in terabytes or even petabytes, with a frequent and rapid rate of growth and change (in some cases, almost in real time). In essence, big data refers to data sets that are too large or too fast-changing to be analyzed using traditional relational or multidimensional database techniques or commonly used software tools to capture, manage and process the data at a reasonable elapsed time.[83]

This change in analytics capabilities dealing with big data can introduce technical and operational risk, and organizations should understand that risk can be incurred either through adoption or non-adoption of these capabilities.

Technical and operational risk should consider that certain data elements may be governed by regulatory or contractual requirements and that data elements may need to be centralized in one place (or at least be accessible centrally) so that the data can be analyzed. In some cases, this centralization can compound technical risk.[84] For example:[85]
- **Amplified technical impact**—If an unauthorized user were to gain access to centralized repositories, it puts the entirety of those data in jeopardy rather than a subset of the data.
- **Privacy (data collection)**—Analytics techniques can impact privacy; for example, individuals whose data are being analyzed may feel that revealed information about them is overly intrusive.
- **Privacy (re-identification)**—Likewise, when data are aggregated, semi-anonymous information or information that is not individually identifiable information might become non-anonymous or identifiable in the process.

---

[83] ISACA, *Big Data: Impacts & Benefits*, USA, 2013, www.isaca.org/Knowledge-Center/Research/Documents/Big-Data_whp_Eng_0413.pdf
[84] ISACA, *Generating Value from Big Data Analytics*, USA, 2014, www.isaca.org/Knowledge-Center/Research/Documents/Generating-Value-from-Big-Data-Analytics_whp_Eng_0114.pdf
[85] *Ibid.*

ARTIFICIAL INTELLIGENCE

Artificial intelligence (AI) is the development of advanced computer systems that can simulate human capabilities, such as analysis. This emerging technology will make it possible for a security team to accurately and efficiently manage huge amounts of complex data and analyze it for new threats.

AI detection schemes will be able to assist with addressing common threats, while also investigating anomalies to detect new and rapidly evolving threats. Furthermore, instead of presenting security analysts with massive amounts of raw data, AI technology can present the information in more practical views and adapt quickly to new methods of attacks.

TOPIC 5—CLOUD AND DIGITAL COLLABORATION

According to NIST and the Cloud Security Alliance (CSA), cloud computing is defined as a "model for enabling convenient, on-demand network access to a shared pool of configurable computing resources (e.g., networks, servers, storage, applications and services) that can be rapidly provisioned and released with minimal management effort or service provider interaction."

Cloud computing offers enterprises a way to save on the capital expenditure associated with traditional methods of managing IT. Common platforms offered in the cloud include Software as a Service (SaaS), Platform as a Service (PaaS) and Infrastructure as a Service (IaaS). Virtualization and service-oriented architectures (SOAs) act as key enablers behind the scenes. Though attractive, cloud computing is not without its own set of risk, first and foremost of which is the safety and security of the data that are entrusted in the care of cloud providers.[86]

Similar to the use of any third-party contract, it is important for organizations to ensure that their cloud provider has a security system in place equivalent to or better than the organization's own security practice. Many cloud providers are ISO 27001 or FIPS 140-2 certified. In addition, organizations can request audits of the cloud provider. The security audits should cover the facilities, networks, hardware and operating systems within the cloud infrastructure.

RISK OF CLOUD COMPUTING

The challenge for cloud computing is to protect data within public and private clouds as well as ensure governance, risk management and compliance are addressed across the full, integrated environment. NIST outlines the following top security risk for cloud infrastructure:

- **Loss of governance**—The client usually relinquishes some level of control to the cloud provider, which may affect security, especially if the service level agreements (SLAs) leave a gap in security defenses.
- **Lock-in**—It can be difficult for a client to migrate from one provider to another, which creates a dependency on a particular cloud provider for service provision.
- **Isolation failure**—One characteristic of cloud computing is shared resources. Although not commonplace, the failure of mechanisms that separate storage, memory, routing and reputation between different tenants can create risk.
- **Compliance**—Migrating to the cloud may create a risk in the organization achieving certification if the cloud provider cannot provide compliance evidence.
- **Management interface compromise**—The customer management interface can pose an increased risk because it is accessed through the Internet and mediates access to larger sets of resources.
- **Data protection**—It may be difficult for clients to check the data handling procedures of the cloud provider.
- **Insecure or incomplete data deletion**—Because of the multiple tenancies and the reuse of hardware resources, there is a greater risk that data are not deleted completely, adequately or in a timely manner.
- **Malicious insider**—Cloud architects have extremely high-risk roles. A malicious insider could cause a great degree of damage.

This risk can lead to a number of different threat events. The CSA lists the following as the top cloud computing threats:[87]
1. Data breaches
2. Data loss
3. Account hijacking
4. Insecure application programming interfaces (APIs)
5. Denial-of-service (DoS)
6. Malicious insiders
7. Abuse of cloud services
8. Insufficient due diligence
9. Shared technology issues

---

[86] ISACA, *Top Business/Technology Issues Survey Results*, USA, 2011
[87] Cloud Security Alliance (CSA), *The Notorious Nine: Cloud Computing Top Threats in 2013*, 2013

CLOUD APPLICATION RISK

In implementing and adapting their cloud-based strategies, enterprises tend to include SaaS offerings, sometimes extending this to critical business processes and related applications. Despite the fact that these service offerings may bring business advantages, they nevertheless generate data-in-flow vulnerabilities that may be exploited by cybercrime and cyberwarfare. The resulting risk is exacerbated by the fact that many vendors and hardware providers for mobile devices, supply cloud-based freeware that is designed to enforce user loyalty. This is often the case for data synchronization, handling of popular file types such as music or pictures, and personal information such as email and calendar entries.

The application layer within the overall IT environment is particularly susceptible to zero-day exploits, as witnessed by many practical examples. Even major software vendors frequently update and patch their applications, but new attack vectors using such applications emerge almost on a daily basis. In terms of cybercrime and cyberwarfare, the market for zero-day exploits is a lively one, and the time span from discovery to recognition and remediation is increasing.

Likewise, the propagation of complex malware has been growing over the past several years. From a cybercrime and cyberwarfare perspective, recent specimens of malware show a higher level of sophistication and persistence than the basic varieties used by opportunistic attackers. While software vendors are quick to address malware in terms of recognition and removal, there is a significant residual risk of malware becoming persistent in target enterprises.

Secondary malware attacks—where APTs make use of already installed simple malware—are often successful where the environmental conditions are conducive to user error or lack of vigilance, namely in home-user or traveling-user scenarios. In practice, removal of the primary malware (a fairly simple process) often allays any further suspicion and causes users and security managers to be lulled into a false sense of security. The secondary and very complex malware may have infiltrated the system, presenting a known and simple piece of primary malware as bait.

BENEFITS OF CLOUD COMPUTING

Although cloud computing is attractive to attackers because of the massive concentrations of data, cloud defenses can be more robust, scalable and cost-effective. ENISA provides the following top security benefits of cloud computing:
- **Market drive**—Because security is a top priority for most cloud customers, cloud providers have a strong driver for increasing and improving their security practices.
- **Scalability**—Cloud technology allows for the rapid reallocation of resources, such as those for filtering, traffic shaping, authentication and encryption, to defensive measures.
- **Cost-effective**—All types of security measures are cheaper when implemented on a large scale. The concentration of resources provides for cheaper physical perimeter and physical access control and easier and cheaper application of many security-related processes.
- **Timely and effective updates**—Updates can be rolled out rapidly across a homogeneous platform.
- **Audit and evidence**—Cloud computing can provide forensic images of virtual machines, which results in less downtime for forensic investigations.

Although there are many benefits of cloud computing, there is risk involved as well. **Figure 6.11** lists the benefits and risk of cloud computing.

Figure 6.11—Benefits and Risk of Cloud Computing

Benefits	Risk
• Market drive for the cloud • Scalability • Cost-effective implementation • Timely and effective updates • Audit and evidence capabilities	• Loss of governance • Lock-in to one provider • Isolation failure • Compliance • Data protection • Customer management interface compromise • Insecure or incomplete data deletion • Malicious insider

SOCIAL MEDIA[88]

Social media technology involves the creation and dissemination of content through social networks using the Internet. The differences between traditional and social media are defined by the level of interaction and interactivity available to the consumer. For example, a viewer can watch the news on television with no interactive feedback mechanisms, while social media tools allow consumers to comment, discuss and even distribute the news. Use of social media has created highly effective communication platforms where any user, virtually anywhere in the world, can freely create content and disseminate this information in real time to a global audience ranging in size from a handful to literally millions.

There are many types of social media tools: blogs (e.g., WordPress), image and video sharing sites (e.g., Flickr and YouTube), social networking (e.g., Facebook), and professional networking (e.g., LinkedIn). The common link among all forms of social media is that the content is supplied and managed by individual users who leverage the tools and platforms provided by social media sites.

Enterprises are using social media to increase brand recognition, sales, revenue and customer satisfaction; however, there is risk associated with its usage. These are divided into those enterprises with a corporate social media presence and those whose employees engage in social media.

Risk associated with a corporate social media presence includes:
• Introduction of viruses/malware to the organizational network
• Misinformation or misleading information posted through a fraudulent or hijacked corporate presence
• Unclear or undefined content rights to information posted to social media sites
• Customer dissatisfaction due to an expected increase in customer service response quality/timeliness
• Mismanagement of electronic communications that may be impacted by retention regulations or ediscovery

Risk associated with employee personal use of social media includes:
• Use of personal accounts to communicate work-related information
• Employee posting of pictures or information that link them to the enterprise
• Excessive employee use of social media in the workplace
• Employee access to social media via enterprise-supplied mobile devices (smartphones, tablets)

---

[88] ISACA, *CISA Review Manual 26th Edition*, USA, 2015

Page intentionally left blank

SECTION 6—KNOWLEDGE CHECK

1. _____ is defined as "a model for enabling convenient, on-demand network access to a shared pool of configurable resources (e.g., networks, servers, storage, applications and services) that can be rapidly provisioned and released with minimal management or service provider interaction."
   A. Software as a Service (SaaS)
   B. Cloud computing
   C. Big data
   D. Platform as a Service (PaaS)

2. Select all that apply. Which of the following statements about advanced persistent threats (APTs) are true?
   A. APTs typically originate from sources such as organized crime groups, activists or governments.
   B. APTs use obfuscation techniques that help them remain undiscovered for months or even years.
   C. APTs are often long-term, multi-phase projects with a focus on reconnaissance.
   D. The APT attack cycle begins with target penetration and collection of sensitive information.
   E. Although they are often associated with APTs, intelligence agencies are rarely the perpetrators of APT attacks.

3. Which of the following are benefits to BYOD?
   A. Acceptable Use Policy is easier to implement.
   B. Costs shift to the user.
   C. Worker satisfaction increases.
   D. Security risk is known to the user.

4. Choose three. Which types of risk are typically associated with mobile devices?
   A. Organizational risk
   B. Compliance risk
   C. Technical risk
   D. Physical risk
   E. Transactional risk

5. Which three elements of the current threat landscape have provided increased levels of access and connectivity, and, therefore, increased opportunities for cybercrime?
   A. Text messaging, Bluetooth technology and SIM cards
   B. Web applications, botnets and primary malware
   C. Financial gains, intellectual property and politics
   D. Cloud computing, social media and mobile computing

See answers in Appendix C.

Page intentionally left blank

Appendices

Appendix A—Knowledge Statements
Appendix B—Glossary
Appendix C—Knowledge Check Answers

Page intentionally left blank

APPENDIX A—KNOWLEDGE STATEMENTS

DOMAIN 1: CYBERSECURITY CONCEPTS

1.1 Knowledge of cybersecurity principles used to manage risk related to the use, processing, storage and transmission of information or data
1.2 Knowledge of security management
1.3 Knowledge of risk management processes, including steps and methods for assessing risk
1.4 Knowledge of threat actors (e.g., script kiddies, non-nation state sponsored, and nation state sponsored)
1.5 Knowledge of cybersecurity roles
1.6 Knowledge of common adversary tactics, techniques, and procedures (TTPs)
1.7 Knowledge of relevant laws, policies, procedures and governance requirements
1.8 Knowledge of cybersecurity controls

DOMAIN 2: CYBERSECURITY ARCHITECTURE PRINCIPLES

2.1 Knowledge of network design processes, to include understanding of security objectives, operational objectives and trade-offs
2.2 Knowledge of security system design methods, tools and techniques
2.3 Knowledge of network access, identity and access management
2.4 Knowledge of information technology (IT) security principles and methods (e.g., firewalls, demilitarized zones, encryption)
2.5 Knowledge of network security architecture concepts, including topology, protocols, components and principles (e.g., application of defense in depth)
2.6 Knowledge of malware analysis concepts and methodology
2.7 Knowledge of intrusion detection methodologies and techniques for detecting host- and network-based intrusions via intrusion detection technologies
2.8 Knowledge of defense in depth principles and network security architecture
2.9 Knowledge of encryption algorithms (e.g., Internet Protocol Security [IPSEC], Advanced Encryption Standard [AES], Generic Routing Encapsulation [GRE])
2.10 Knowledge of cryptography
2.11 Knowledge of encryption methodologies
2.12 Knowledge of how traffic flows across the network (i.e., transmission and encapsulation)
2.13 Knowledge of network protocols (e.g., Transmission Control Protocol and Internet Protocol [TCP/IP], Dynamic Host Configuration Protocol [DHCP]), and directory services (e.g., domain name system [DNS])

DOMAIN 3: SECURITY OF NETWORK, SYSTEM, APPLICATION AND DATA

3.1 Knowledge of vulnerability assessment tools, including open source tools, and their capabilities
3.2 Knowledge of basic system administration, network and operating system hardening techniques.
3.3 Knowledge of risk associated with virtualizations
3.4 Knowledge of penetration testing
3.5 Knowledge of network systems management principles, models, methods (e.g., end-to-end systems performance monitoring) and tools
3.6 Knowledge of remote access technology
3.7 Knowledge of UNIX command line
3.8 Knowledge of system and application security threats and vulnerabilities
3.9 Knowledge of system life cycle management principles, including software security and usability
3.10 Knowledge of local specialized system requirements (e.g., critical infrastructure systems that may not use standard information technology [IT]) for safety, performance and reliability
3.11 Knowledge of system and application security threats and vulnerabilities (e.g., buffer overflow, mobile code, cross-site scripting, Procedural Language/Structured Query Language [PL/SQL] and injections, race conditions, cover channel, replay, return-oriented attacks, malicious code)
3.12 Knowledge of social dynamics of computer attackers in a global context
3.13 Knowledge of secure configuration management techniques

3.14 Knowledge of capabilities and applications of network equipment including hubs, routers, switches, bridges, servers, transmission media and related hardware
3.15 Knowledge of communication methods, principles and concepts that support the network infrastructure
3.16 Knowledge of the common networking protocols (e.g., Transmission Control Protocol and Internet Protocol [TCP/IP]) and services (e.g., web, email, domain name system [DNS]) and how they interact to provide network communications
3.17 Knowledge of different types of network communication (e.g., local area network [LAN], wide area network [WAN], metropolitan area network [MAN], wireless local area network [WLAN], wireless wide area network [WWAN])
3.18 Knowledge of virtualization technologies and virtual machine development and maintenance
3.19 Knowledge of application security (e.g., system development life cycle [SDLC], vulnerabilities, best practices)
3.20 Knowledge of risk threat assessment

DOMAIN 4: INCIDENT RESPONSE
4.1   Knowledge of incident categories for responses
4.2   Knowledge of business continuity/disaster recovery
4.3   Knowledge of incident response and handling methodologies
4.4   Knowledge of security event correlation tools
4.5   Knowledge of processes for seizing and preserving digital evidence (e.g., chain of custody)
4.6   Knowledge of types of digital forensics data
4.7   Knowledge of basic concepts and practices of processing digital forensic data
4.8   Knowledge of anti-forensics tactics, techniques and procedures (TTPS)
4.9   Knowledge of common forensic tool configuration and support applications (e.g., VMware®, Wireshark®)
4.10 Knowledge of network traffic analysis methods
4.11 Knowledge of which system files (e.g., log files, registry files, configuration files) contain relevant information and where to find those system files

DOMAIN 5: SECURITY OF EVOLVING TECHNOLOGY
5.1   Knowledge of emerging technology and associated security issues, risk and vulnerabilities
5.2   Knowledge of risk associated with mobile computing
5.3   Knowledge of cloud concepts around data and collaboration
5.4   Knowledge of risk of moving applications and infrastructure to the cloud
5.5   Knowledge of risk associated with outsourcing
5.6   Knowledge of supply chain risk management processes and practices

APPENDIX B—GLOSSARY

A

Acceptable interruption window—The maximum period of time that a system can be unavailable before compromising the achievement of the enterprise's business objectives.

Acceptable use policy—A policy that establishes an agreement between users and the enterprise and defines for all parties' the ranges of use that are approved before gaining access to a network or the Internet.

Access control list (ACl)—An internal computerized table of access rules regarding the levels of computer access permitted to logon IDs and computer terminals. Also referred to as access control tables.

Access path—The logical route that an end user takes to access computerized information. Typically includes a route through the operating system, telecommunications software, selected application software and the access control system.

Access rights—The permission or privileges granted to users, programs or workstations to create, change, delete or view data and files within a system, as defined by rules established by data owners and the information security policy.

Accountability—The ability to map a given activity or event back to the responsible party.

Advanced Encryption Standard (AES)—A public algorithm that supports keys from 128 bits to 256 bits in size.

Advanced persistent threat (APT)—An adversary that possesses sophisticated levels of expertise and significant resources which allow it to create opportunities to achieve its objectives using multiple attack vectors (NIST SP800-61).

The APT:
1. Pursues its objectives repeatedly over an extended period of time
2. Adapts to defenders' efforts to resist it
3. Is determined to maintain the level of interaction needed to execute its objectives

Adversary—A threat agent.

Adware— A software package that automatically plays, displays or downloads advertising material to a computer after the software is installed on it or while the application is being used. In most cases, this is done without any notification to the user or without the user's consent. The term adware may also refer to software that displays advertisements, whether or not it does so with the user's consent; such programs display advertisements as an alternative to shareware registration fees. These are classified as adware in the sense of advertising supported software, but not as spyware. Adware in this form does not operate surreptitiously or mislead the user, and it provides the user with a specific service.

Alert situation—The point in an emergency procedure when the elapsed time passes a threshold and the interruption is not resolved. The enterprise entering into an alert situation initiates a series of escalation steps.

Alternate facilities—Locations and infrastructures from which emergency or backup processes are executed, when the main premises are unavailable or destroyed; includes other buildings, offices or data processing centers.

Alternate process—Automatic or manual process designed and established to continue critical business processes from point-of-failure to return-to- normal.

Analog—A transmission signal that varies continuously in amplitude and time and is generated in wave formation. Analog signals are used in telecommunications.

Anti-malware—A technology widely used to prevent, detect and remove many categories of malware, including computer viruses, worms, Trojans, keyloggers, malicious browser plug-ins, adware and spyware.

Antivirus software—An application software deployed at multiple points in an IT architecture. It is designed to detect and potentially eliminate virus code before damage is done and repair or quarantine files that have already been infected.

Application layer—In the Open Systems Interconnection (OSI) communications model, the application layer provides services for an application program to ensure that effective communication with another application program in a network is possible. The application layer is not the application that is doing the communication; a service layer that provides these services.

Architecture—Description of the fundamental underlying design of the components of the business system, or of one element of the business system (e.g., technology), the relationships among them, and the manner in which they support enterprise objectives.

Asset—Something of either tangible or intangible value that is worth protecting, including people, information, infrastructure, finances and reputation.

Asymmetric key (public key)—A cipher technique in which different cryptographic keys are used to encrypt and decrypt a message. See public key encryption.

Attack—An actual occurrence of an adverse event.

Attack mechanism—A method used to deliver the payload. Unless the attacker is personally performing the attack, an attack mechanism may involve an exploit delivering a payload to the target.

Attack vector—A path or route used by the adversary to gain access to the target (asset). There are two types of attack vectors: ingress and egress (also known as data exfiltration).

Attenuation—Reduction of signal strength during transmission.

Audit trail—A visible trail of evidence enabling one to trace information contained in statements or reports back to the original input source.

Authentication—The act of verifying the identity of a user and the user's eligibility to access computerized information. Authentication is designed to protect against fraudulent logon activity. It can also refer to the verification of the correctness of a piece of data.

Authenticity—Undisputed authorship.

Availability—Ensuring timely and reliable access to and use of information.

B
Back door—A means of regaining access to a compromised system by installing software or configuring existing software to enable remote access under attacker-defined conditions.

Bandwidth—The range between the highest and lowest transmittable frequencies. It equates to the transmission capacity of an electronic line and is expressed in bytes per second or Hertz (cycles per second).

Bastion—System heavily fortified against attacks.

Biometrics—A security technique that verifies an individual's identity by analyzing a unique physical attribute, such as a handprint.

Block cipher—A public algorithm that operates on plaintext in blocks (strings or groups) of bits.

Botnet—A term derived from "robot network;" is a large automated and distributed network of previously compromised computers that can be simultaneously controlled to launch large-scale attacks such as a denial-of-service attack on selected victims.

Boundary—Logical and physical controls to define a perimeter between the organization and the outside world.

Bridges—Data link layer devices developed in the early 1980s to connect local area networks (LANs) or create two separate LAN or wide area network (WAN) network segments from a single segment to reduce collision domains. Bridges act as store- and-forward devices in moving frames toward their destination. This is achieved by analyzing the MAC header of a data packet, which represents the hardware address of an NIC.

Bring your own device (BYOD)—An enterprise policy used to permit partial or full integration of user-owned mobile devices for business purposes.

Broadcast—A method to distribute information to multiple recipients simultaneously.

Brute force—A class of algorithms that repeatedly try all possible combinations until a solution is found.

Brute force attack—Repeatedly trying all possible combinations of passwords or encryption keys until the correct one is found.

Buffer overflow—Occurs when a program or process tries to store more data in a buffer (temporary data storage area) than it was intended to hold. Since buffers are created to contain a finite amount of data, the extra information—which has to go somewhere—can overflow into adjacent buffers, corrupting or overwriting the valid data held in them. Although it may occur accidentally through programming error, buffer overflow is an increasingly common type of security attack on data integrity. In buffer overflow attacks, the extra data may contain codes designed to trigger specific actions, in effect sending new instructions to the attacked computer that could, for example, damage the user's files, change data, or disclose confidential information. Buffer overflow attacks are said to have arisen because the C programming language supplied the framework, and poor programming practices supplied the vulnerability.

Business continuity plan (BCP)—A plan used by an enterprise to respond to disruption of critical business processes. Depends on the contingency plan for restoration of critical systems.

Business impact analysis/assessment (BIA)— Evaluating the criticality and sensitivity of information assets. An exercise that determines the impact of losing the support of any resource to an enterprise, establishes the escalation of that loss over time, identifies the minimum resources needed to recover, and prioritizes the recovery of processes and the supporting system. This process also includes addressing income loss, unexpected expense, legal issues (regulatory compliance or contractual), interdependent processes, and loss of public reputation or public confidence.

C

Certificate (Certification) authority (CA)—A trusted third party that serves authentication infrastructures or enterprises and registers entities and issues them certificates.

Certificate revocation list (CRL)—An instrument for checking the continued validity of the certificates for which the certification authority (CA) has responsibility. The CRL details digital certificates that are no longer valid. The time gap between two updates is very critical and is also a risk in digital certificates verification.

Appendix B—Glossary

Chain of custody—A legal principle regarding the validity and integrity of evidence. It requires accountability for anything that will be used as evidence in a legal proceeding to ensure that it can be accounted for from the time it was collected until the time it is presented in a court of law. Includes documentation as to who had access to the evidence and when, as well as the ability to identify evidence as being the exact item that was recovered or tested. Lack of control over evidence can lead to it being discredited. Chain of custody depends on the ability to verify that evidence could not have been tampered with. This is accomplished by sealing off the evidence, so it cannot be changed, and providing a documentary record of custody to prove that the evidence was at all times under strict control and not subject to tampering.

Checksum— A mathematical value that is assigned to a file and used to "test" the file at a later date to verify that the data contained in the file has not been maliciously changed. A cryptographic checksum is created by performing a complicated series of mathematical operations (known as a cryptographic algorithm) that translates the data in the file into a fixed string of digits called a hash value, which is then used as the checksum. Without knowing which cryptographic algorithm was used to create the hash value, it is highly unlikely that an unauthorized person would be able to change data without inadvertently changing the corresponding checksum. Cryptographic checksums are used in data transmission and data storage. Cryptographic checksums are also known as message authentication codes, integrity check-values, modification detection codes or message integrity codes.

Chief Information Security officer (CISO)—The person in charge of information security within the enterprise.

Chief Security officer (CSO)—The person usually responsible for all security matters both physical and digital in an enterprise.

Cipher—An algorithm to perform encryption.

Ciphertext—Information generated by an encryption algorithm to protect the plaintext and that is unintelligible to the unauthorized reader.

Cleartext—Data that is not encrypted. Also known as plaintext.

Cloud computing—Convenient, on-demand network access to a shared pool of resources that can be rapidly provisioned and released with minimal management effort or service provider interaction.

Collision—The situation that occurs when two or more demands are made simultaneously on equipment that can handle only one at any given instant (Federal Standard 1037C).

Common attack pattern enumeration and classification (CAPEC)—A catalogue of attack patterns as "an abstraction mechanism for helping describe how an attack against vulnerable systems or networks is executed" published by the MITRE Corporation.

Compartmentalization—A process for protecting very high value assets or in environments where trust is an issue. Access to an asset requires two or more processes, controls or individuals.

Compliance—Adherence to, and the ability to demonstrate adherence to, mandated requirements defined by laws and regulations, as well as voluntary requirements resulting from contractual obligations and internal policies.

Compliance documents—Policies, standard and procedures that document the actions that are required or prohibited. Violations may be subject to disciplinary actions.

Computer emergency response team (CERT)—A group of people integrated at the enterprise with clear lines of reporting and responsibilities for standby support in case of an information systems emergency. This group will act as an efficient corrective control, and should also act as a single point of contact for all incidents and issues related to information systems.

Computer forensics—The application of the scientific method to digital media to establish factual information for judicial review. This process often involves investigating computer systems to determine whether they are or have been used for illegal or unauthorized activities. As a discipline, it combines elements of law and computer science to collect and analyze data from information systems (e.g., personal computers, networks, wireless communication and digital storage devices) in a way that is admissible as evidence in a court of law.

Confidentiality—Preserving authorized restrictions on access and disclosure, including means for protecting privacy and proprietary information.

Configuration management—The control of changes to a set of configuration items over a system life cycle.

Consumerization—A new model in which emerging technologies are first embraced by the consumer market and later spread to the business.

Containment—Actions taken to limit exposure after an incident has been identified and confirmed.

Content filtering—Controlling access to a network by analyzing the contents of the incoming and outgoing packets and either letting them pass or denying them based on a list of rules. Differs from packet filtering in that it is the data in the packet that are analyzed instead of the attributes of the packet itself (e.g., source/target IP address, transmission control protocol [TCP] flags).

Control—The means of managing risk, including policies, procedures, guidelines, practices or organizational structures, which can be of an administrative, technical, management, or legal nature. Also used as a synonym for safeguard or countermeasure.

Countermeasure—Any process that directly reduces a threat or vulnerability.

Critical infrastructure—Systems whose incapacity or destruction would have a debilitating effect on the economic security of an enterprise, community or nation.

Criticality—The importance of a particular asset or function to the enterprise, and the impact if that asset or function is not available.

Criticality analysis—An analysis to evaluate resources or business functions to identify their importance to the enterprise, and the impact if a function cannot be completed or a resource is not available.

Cross-site scripting (XSS)—A type of injection, in which malicious scripts are injected into otherwise benign and trusted websites. Cross-site scripting (XSS) attacks occur when an attacker uses a web application to send malicious code, generally in the form of a browser side script, to a different end user. Flaws that allow these attacks to succeed are quite widespread and occur anywhere a web application uses input from a user within the output it generates without validating or encoding it. (OWASP)

Cryptography—The art of designing, analyzing and attacking cryptographic schemes.

Cryptosystem—A pair of algorithms that take a key and convert plaintext to ciphertext and back.

Cybercop—An investigator of activities related to computer crime.

Cyberespionage—Activities conducted in the name of security, business, politics or technology to find information that ought to remain secret. It is not inherently military.

Cybersecurity—The protection of information assets by addressing threats to information processed, stored, and transported by internetworked information systems.

Appendix B—Glossary

Cybersecurity architecture—Describes the structure, components and topology (connections and layout) of security controls within an enterprise's IT infrastructure. The security architecture shows how defense in depth is implemented and how layers of control are linked and is essential to designing and implementing security controls in any complex environment.

Cyberwarfare—Activities supported by military organizations with the purpose to threat the survival and well-being of society/foreign entity.

D

Data classification—The assignment of a level of sensitivity to data (or information) that results in the specification of controls for each level of classification. Levels of sensitivity of data are assigned according to predefined categories as data are created, amended, enhanced, stored or transmitted. The classification level is an indication of the value or importance of the data to the enterprise.

Data custodian—The individual(s) and department(s) responsible for the storage and safeguarding of computerized data.

Data Encryption Standard (DES)—An algorithm for encoding binary data. It is a secret key cryptosystem published by the National Bureau of Standards (NBS), the predecessor of the US National Institute of Standards and Technology (NIST). DES and its variants has been replaced by the Advanced Encryption Standard (AES).

Data leakage—Siphoning out or leaking information by dumping computer files or stealing computer reports and tapes.

Data owner—The individual(s), normally a manager or director, who has responsibility for the integrity, accurate reporting and use of computerized data.

Data retention—Refers to the policies that govern data and records management for meeting internal, legal and regulatory data archival requirements.

Database—A stored collection of related data needed by enterprises and individuals to meet their information processing and retrieval requirements.

Decentralization—The process of distributing computer processing to different locations within an enterprise.

Decryption—A technique used to recover the original plaintext from the ciphertext so that it is intelligible to the reader. The decryption is a reverse process of the encryption.

Decryption key—A digital piece of information used to recover plaintext from the corresponding ciphertext by decryption.

Defense in depth—The practice of layering defenses to provide added protection. Defense in depth increases security by raising the effort needed in an attack. This strategy places multiple barriers between an attacker and an enterprise's computing and information resources.

Demilitarized zone (DMZ)—A screened (firewalled) network segment that acts as a buffer zone between a trusted and untrusted network. A DMZ is typically used to house systems such as web servers that must be accessible from both internal networks and the Internet.

Denial-of-service (DoS) attack—An assault on a service from a single source that floods it with so many requests that it becomes overwhelmed and is either stopped completely or operates at a significantly reduced rate.

Digital certificate— A piece of information, a digitized form of signature, that provides sender authenticity, message integrity and nonrepudiation. A digital signature is generated using the sender's private key or applying a one-way hash function.

Digital forensics—The process of identifying, preserving, analyzing and presenting digital evidence in a manner that is legally acceptable in any legal proceedings.

Digital signature—A piece of information, a digitized form of signature, that provides sender authenticity, message integrity and nonrepudiation. A digital signature is generated using the sender's private key or applying a one-way hash function.

Disaster—A sudden, unplanned calamitous event causing great damage or loss. Any event that creates an inability on an organization's part to provide critical business functions for some predetermined period of time. Similar terms are business interruption, outage and catastrophe.

The period when enterprise management decides to divert from normal production responses and exercises its disaster recovery plan (DRP). It typically signifies the beginning of a move from a primary location to an alternate location.

Disaster recovery plan (DRP)—A set of human, physical, technical and procedural resources to recover, within a defined time and cost, an activity interrupted by an emergency or disaster.

Discretionary access control (DAC)—A means of restricting access to objects based on the identity of subjects and/or groups to which they belong. The controls are discretionary in the sense that a subject with a certain access permission is capable of passing that permission (perhaps indirectly) on to any other subject.

Domain name system (DNS)—A hierarchical database that is distributed across the Internet that allows names to be resolved into IP addresses (and vice versa) to locate services such as web and email servers.

Domain name system (DNS) exfiltration— Tunneling over DNS to gain network access. Lower-level attack vector for simple to complex data transmission, slow but difficult to detect.

Due care—The level of care expected from a reasonable person of similar competency under similar conditions.

Due diligence—The performance of those actions that are generally regarded as prudent, responsible and necessary to conduct a thorough and objective investigation, review and/or analysis.

Dynamic ports—Dynamic and/or private ports--49152 through 65535: Not listed by IANA because of their dynamic nature.

E

Eavesdropping—Listening a private communication without permission.

Ecommerce—The processes by which enterprises conduct business electronically with their customers, suppliers and other external business partners, using the Internet as an enabling technology. Ecommerce encompasses both business-to- business (B2B) and business-to-consumer (B2C) ecommerce models, but does not include existing non-Internet ecommerce methods based on private networks such as electronic data interchange (EDI) and Society for Worldwide Interbank Financial Telecommunication (SWIFT).

Egress—Network communications going out.

Elliptical curve cryptography (ECC)—An algorithm that combines plane geometry with algebra to achieve stronger authentication with smaller keys compared to traditional methods, such as RSA, which primarily use algebraic factoring. Smaller keys are more suitable to mobile devices.

Encapsulation security payload (ESP)—Protocol, which is designed to provide a mix of security services in IPv4 and IPv6. ESP can be used to provide confidentiality, data origin authentication, connectionless integrity, an anti-replay service (a form of partial sequence integrity), and (limited) traffic flow confidentiality (RFC 4303). The ESP header is inserted after the IP header and before the next layer protocol header (transport mode) or before an encapsulated IP header (tunnel mode).

Encryption—The process of taking an unencrypted message (plaintext), applying a mathematical function to it (encryption algorithm with a key) and producing an encrypted message (ciphertext).

Encryption algorithm—A mathematically based function or calculation that encrypts/decrypts data.

Encryption key—A piece of information, in a digitized form, used by an encryption algorithm to convert the plaintext to the ciphertext.

Eradication—When containment measures have been deployed after an incident occurs, the root cause of the incident must be identified and removed from the network. Eradication methods include: restoring backups to achieve a clean state of the system, removing the root cause, improving defenses and performing vulnerability analysis to find further potential damage from the same root cause.

Ethernet—A popular network protocol and cabling scheme that uses a bus topology and carrier sense multiple access/collision detection (CSMA/CD) to prevent network failures or collisions when two devices try to access the network at the same time.

Event—Something that happens at a specific place and/or time.

Evidence—Information that proves or disproves a stated issue. Information that an auditor gathers in the course of performing an IS audit; relevant if it pertains to the audit objectives and has a logical relationship to the findings and conclusions it is used to support.

Exploit—Full use of a vulnerability for the benefit of an attacker.

F

File transfer protocol (FTP)—A protocol used to transfer files over a Transmission Control Protocol/ Internet Protocol (TCP/IP) network (Internet, UNIX, etc.).

Firewall—A system or combination of systems that enforces a boundary between two or more networks, typically forming a barrier between a secure and an open environment such as the Internet.

Forensic examination—The process of collecting, assessing, classifying and documenting digital evidence to assist in the identification of an offender and the method of compromise.

Freeware—Software available free of charge.

G

Gateway—A device (router, firewall) on a network that serves as an entrance to another network.

Governance—Ensures that stakeholder needs, conditions and options are evaluated to determine balanced, agreed-on enterprise objectives to be achieved; setting direction through prioritization and decision making; and monitoring performance and compliance against agreed-on direction and objectives. Conditions can include the cost of capital, foreign exchange rates, etc. Options can include shifting manufacturing to other locations, subcontracting portions of the enterprise to third parties, selecting a product mix from many available choices, etc.

Governance, Risk Management and Compliance (GRC)—A business term used to group the three close-related disciplines responsible for the protection of assets and operations.

Guideline—A description of a particular way of accomplishing something that is less prescriptive than a procedure.

H

Hacker—An individual who attempts to gain unauthorized access to a computer system.

Hash function—An algorithm that maps or translates one set of bits into another (generally smaller) so that a message yields the same result every time the algorithm is executed using the same message as input. It is computationally infeasible for a message to be derived or reconstituted from the result produced by the algorithm or to find two different messages that produce the same hash result using the same algorithm.

Hash total—The total of any numeric data field in a document or computer file. This total is checked against a control total of the same field to facilitate accuracy of processing.

Hashing—Using a hash function (algorithm) to create hash valued or checksums that validate message integrity.

Hijacking—An exploitation of a valid network session for unauthorized purposes.

Honeypot—A specially configured server, also known as a decoy server, designed to attract and monitor intruders in a manner such that their actions do not affect production systems. Also known as "decoy server."

Horizontal defense in depth—Controls are placed in various places in the path to access an asset.

Hubs—A common connection point for devices in a network, hubs are used to connect segments of a local area network (LAN). A hub contains multiple ports. When a packet arrives at one port, it is copied to the other ports so that all segments of the LAN can see all packets.

Human firewall—A person prepared to act as a network layer of defense through education and awareness.

Hypertext Transfer protocol (HTTP)—A communication protocol used to connect to servers on the World Wide Web. Its primary function is to establish a connection with a web server and transmit hypertext markup language (HTML), extensible markup language (XML) or other pages to client browsers.

I

IEEE (Institute of Electrical and Electronics Engineers)—Pronounced I-triple-E; an organization composed of engineers, scientists and students. Best known for developing standards for the computer and electronics industry.

IEEE 802.11—A family of specifications developed by the Institute of Electrical and Electronics Engineers (IEEE) for wireless local area network (WLAN) technology. 802.11 specifies an over-the- air interface between a wireless client and a base station or between two wireless clients.

Imaging—A process that allows one to obtain a bit-for-bit copy of data to avoid damage of original data or information when multiple analyses may be performed. The imaging process is made to obtain residual data, such as deleted files, fragments of deleted files and other information present, from the disk for analysis. This is possible because imaging duplicates the disk surface, sector by sector.

Impact—Magnitude of loss resulting from a threat exploiting a vulnerability.

Impact analysis—A study to prioritize the criticality of information resources for the enterprise based on costs (or consequences) of adverse events. In an impact analysis, threats to assets are identified and potential business losses determined for different time periods. This assessment is used to justify the extent of safeguards that are required and recovery time frames. This analysis is the basis for establishing the recovery strategy.

Incident—Any event that is not part of the standard operation of a service and that causes, or may cause, an interruption to, or a reduction in, the quality of that service.

Incident response—The response of an enterprise to a disaster or other significant event that may significantly affect the enterprise, its people, or its ability to function productively. An incident response may include evacuation of a facility, initiating a disaster recovery plan (DRP), performing damage assessment, and any other measures necessary to bring an enterprise to a more stable status.

Incident response plan—The operational component of incident management. The plan includes documented procedures and guidelines for defining the criticality of incidents, reporting and escalation process, and recovery procedures.

Information security—Ensures that within the enterprise, information is protected against disclosure to unauthorized users (confidentiality), improper modification (integrity), and nonaccess when required (availability).

Information security program—The overall combination of technical, operational and procedural measures and management structures implemented to provide for the confidentiality, integrity and availability of information based on business requirements and risk analysis.

Information systems (IS)—The combination of strategic, managerial and operational activities involved in gathering, processing, storing, distributing and using information and its related technologies.

Information systems are distinct from information technology (IT) in that an information system has an IT component that interacts with the process components.

Infrastructure as a Service (IaaS)—Offers the capability to provision processing, storage, networks and other fundamental computing resources, enabling the customer to deploy and run arbitrary software, which can include operating systems (OSs) and applications.

Ingestion—A process to convert information extracted to a format that can be understood by investigators. See also Normalization.

Ingress—Network communications coming in.

Inherent risk—The risk level or exposure without taking into account the actions that management has taken or might take (e.g., implementing controls).

Injection—A general term for attack types which consist of injecting code that is then interpreted/ executed by the application (OWASP).

Intangible asset—An asset that is not physical in nature. Examples include: intellectual property (patents, trademarks, copyrights, processes), goodwill and brand recognition.

Integrity—The guarding against improper information modification or destruction, and includes ensuring information nonrepudiation and authenticity.

Intellectual property—Intangible assets that belong to an enterprise for its exclusive use. Examples include: patents, copyrights, trademarks, ideas, and trade secrets.

International Standards organization (ISO)—The world's largest developer of voluntary International Standards.

Internet Assigned Numbers Authority (IANA)— Responsible for the global coordination of the DNS root, IP addressing, and other Internet protocol resources.

Internet Control Message Protocol (ICMP)—A set of protocols that allow systems to communicate information about the state of services on other systems. For example, ICMP is used in determining whether systems are up, maximum packet sizes on links, whether a destination host/network/port is available. Hackers typically use (abuse) ICMP to determine information about the remote site.

Internet protocol (IP)—Specifies the format of packets and the addressing scheme.

Internet protocol (IP) packet spoofing—An attack using packets with the spoofed source Internet packet (IP) addresses. This technique exploits applications that use authentication based on IP addresses. This technique also may enable an unauthorized user to gain root access on the target system.

Internet service provider (ISP)—A third party that provides individuals and enterprises with access to the Internet and a variety of other Internet-related services.

Internetwork packet Exchange/Sequenced packet Exchange (IPX/SPX)—IPX is Layer 3 of the open systems interconnect (OSI) model network protocol; SPX is Layer 4 transport protocol. The SPX layer sits on top of the IPX layer and provides connection- oriented services between two nodes on the network.

Interrogation—Used to obtain prior indicators or relationships, including telephone numbers, IP addresses and names of individuals, from extracted data.

Intruder—Individual or group gaining access to the network and its resources without permission.

Intrusion detection—The process of monitoring the events occurring in a computer system or network to detect signs of unauthorized access or attack.

Intrusion detection system (IDS)—Inspects network and host security activity to identify suspicious patterns that may indicate a network or system attack.

Intrusion prevention—A preemptive approach to network security used to identify potential threats and respond to them to stop, or at least limit, damage or disruption.

Intrusion prevention system (IPS)—A system designed to not only detect attacks, but also to prevent the intended victim hosts from being affected by the attacks.

Investigation—The collection and analysis of evidence with the goal to identifying the perpetrator of an attack or unauthorized use or access.

IP address—A unique binary number used to identify devices on a TCP/IP network.

IP Authentication header (Ah)—Protocol used to provide connectionless integrity and data origin authentication for IP datagrams (hereafter referred to as just "integrity") and to provide protection against replays. (RFC 4302). AH ensures data integrity with a checksum that a message authentication code, such as MD5, generates. To ensure data origin authentication, AH includes a secret shared key in the algorithm that it uses for authentication. To ensure replay protection, AH uses a sequence number field within the IP authentication header.

IP Security (IpSec)—A set of protocols developed by the Internet Engineering Task Force (IETF) to support the secure exchange of packets.

IT governance—The responsibility of executives and the board of directors; consists of the leadership, organizational structures and processes that ensure that the enterprise's IT sustains and extends the enterprise's strategies and objectives.

K

Kernel mode—Used for execution of privileged instructions for the internal operation of the system. In kernel mode, there are no protections from errors or malicious activity and all parts of the system and memory are accessible.

Key length—The size of the encryption key measured in bits.

Key risk indicator (KRI)—A subset of risk indicators that are highly relevant and possess a high probability of predicting or indicating important risk. See also Risk Indicator.

Keylogger—Software used to record all keystrokes on a computer.

L

Latency—The time it takes a system and network delay to respond. More specifically, system latency is the time a system takes to retrieve data. Network latency is the time it takes for a packet to travel from source to the final destination.

Layer 2 switches—Data link level devices that can divide and interconnect network segments and help to reduce collision domains in Ethernet-based networks.

Layer 3 and 4 switches—Switches with operating capabilities at Layer 3 and Layer 4 of the open systems interconnect (OSI) model. These switches look at the incoming packet's networking protocol, e.g., IP, and then compare the destination IP address to the list of addresses in their tables, to actively calculate the best way to send a packet to its destination.

Layer 4-7 switches—Used for load balancing among groups of servers. Also known as content- switches, content services switches, web-switches or application-switches.

Legacy system—Outdated computer systems.

Likelihood—The probability of something happening.

Local area network (LAN)—Communication network that serves several users within a specified geographic area. A personal computer LAN functions as a distributed processing system in which each computer in the network does its own processing and manages some of its data. Shared data are stored in a file server that acts as a remote disk drive for all users in the network.

Log—To record details of information or events in an organized record-keeping system, usually sequenced in the order in which they occurred.

Logical access—Ability to interact with computer resources granted using identification, authentication and authorization.

Logical access controls—The policies, procedures, organizational structure and electronic access controls designed to restrict access to computer software and data files.

M

Media access control (MAC) address—A unique identifier assigned to network interfaces for communications on the physical network segment.

MAC header—Represents the hardware address of an network interface controller (NIC) inside a data packet.

Mail relay server—An electronic mail (email) server that relays messages so that neither the sender nor the recipient is a local user.

Mainframe—A large high-speed computer, especially one supporting numerous workstations or peripherals.

Malware—Short for malicious software. Designed to infiltrate, damage or obtain information from a computer system without the owner's consent. Malware is commonly taken to include computer viruses, worms, Trojan horses, spyware and adware. Spyware is generally used for marketing purposes and, as such, is not malicious, although it is generally unwanted. Spyware can, however, be used to gather information for identity theft or other clearly illicit purposes.

Mandatory access control (MAC)—A means of restricting access to data based on varying degrees of security requirements for information contained in the objects and the corresponding security clearance of users or programs acting on their behalf.

Man-in-the-middle attack—An attack strategy in which the attacker intercepts the communication stream between two parts of the victim system and then replaces the traffic between the two components with the intruder's own, eventually assuming control of the communication.

Masking—A computerized technique of blocking out the display of sensitive information, such as passwords, on a computer terminal or report.

Message authentication code—An American National Standards Institute (ANSI) standard checksum that is computed using Data Encryption Standard (DES).

Message digest—A smaller extrapolated version of the original message created using a message digest algorithm.

Message digest algorithm—Message digest algorithms are SHA1, MD2, MD4 and MD5. These algorithms are one-way functions unlike private and public key encryption algorithms. All digest algorithms take a message of arbitrary length and produce a 128-bit message digest.

Metropolitan area network (MAN)—A data network intended to serve an area the size of a large city.

Miniature fragment attack—Using this method, an attacker fragments the IP packet into smaller ones and pushes it through the firewall, in the hope that only the first of the sequence of fragmented packets would be examined and the others would pass without review.

Mirrored site—An alternate site that contains the same information as the original. Mirrored sites are set up for backup and disaster recovery and to balance the traffic load for numerous download requests. Such download mirrors are often placed in different locations throughout the Internet.

Mobile device—A small, handheld computing devices, typically having a display screen with touch input and/or a miniature keyboard and weighing less than two pounds.

Mobile site—The use of a mobile/temporary facility to serve as a business resumption location. The facility can usually be delivered to any site and can house information technology and staff.

Monitoring policy—Rules outlining or delineating the way in which information about the use of computers, networks, applications and information is captured and interpreted.

Multifactor authentication—A combination of more than one authentication method, such as token and password (or personal identification number [PIN] or token and biometric device).

Appendix B—Glossary

N

National Institute for Standards and Technology (NIST)—Develops tests, test methods, reference data, proof-of concept implementations, and technical analyses to advance the development and productive use of information technology. NIST is a US government entity that creates mandatory standards that are followed by federal agencies and those doing business with them.

Network basic input/output system (NetBIOS)—A program that allows applications on different computers to communicate within a local area network (LAN).

Network address translation (NAT)—A methodology of modifying network address information in datagram packet headers while they are in transit across a traffic routing device for the purpose of remapping one IP address space into another.

Network interface card (NIC)—A communication card that when inserted into a computer, allows it to communicate with other computers on a network. Most NICs are designed for a particular type of network or protocol.

Network news transfer protocol (NNTP)—Used for the distribution, inquiry, retrieval, and posting of Netnews articles using a reliable stream-based mechanism. For news-reading clients, NNTP enables retrieval of news articles that are stored in a central database, giving subscribers the ability to select only those articles they wish to read (RFC 3977).

Network segmentation—A common technique to implement network security is to segment an organization's network into separate zones that can be separately controlled, monitored and protected.

Network traffic analysis—Identifies patterns in network communications. Traffic analysis does not need to have the actual content of the communication but analyzes where traffic is taking place, when and for how long communications occur and the size of information transferred.

Nonintrusive monitoring—The use of transported probes or traces to assemble information, track traffic and identify vulnerabilities.

Nonrepudiation—The assurance that a party cannot later deny originating data; provision of proof of the integrity and origin of the data and that can be verified by a third party. A digital signature can provide nonrepudiation.

Normalization—The elimination of redundant data.

O

Obfuscation—The deliberate act of creating source or machine code that is difficult for humans to understand.

Open Systems Interconnect (OSI) model—A model for the design of a network. The open systems interconnect (OSI) model defines groups of functionality required to network computers into layers. Each layer implements a standard protocol to implement its functionality. There are seven layers in the OSI model.

Operating system (OS)—A master control program that runs the computer and acts as a scheduler and traffic controller.

Open Web Application Security Project (OWASP)—An open community dedicated to enabling organizations to conceive, develop, acquire, operate, and maintain applications that can be trusted.

Outcome measure—Represents the consequences of actions previously taken; often referred to as a lag indicator. Outcome measures frequently focus on results at the end of a time period and characterize historic performance. They are also referred to as a key goal indicator (KGI) and used to indicate whether goals have been met. These can be measured only after the fact and, therefore, are called "lag indicators."

Outsourcing—A formal agreement with a third party to perform IS or other business functions for an enterprise.

P

Packet—Data unit that is routed from source to destination in a packet-switched network. A packet contains both routing information and data. Transmission Control Protocol/Internet Protocol (TCP/IP) is such a packet-switched network.

Packet filtering—Controlling access to a network by analyzing the attributes of the incoming and outgoing packets and either letting them pass, or denying them, based on a list of rules.

Packet switching—The process of transmitting messages in convenient pieces that can be reassembled at the destination.

Passive response—A response option in intrusion detection in which the system simply reports and records the problem detected, relying on the user to take subsequent action.

Password—A protected, generally computer-encrypted string of characters that authenticate a computer user to the computer system.

Password cracker—A tool that tests the strength of user passwords by searching for passwords that are easy to guess. It repeatedly tries words from specially crafted dictionaries and often also generates thousands (and in some cases, even millions) of permutations of characters, numbers and symbols.

Patch—Fixes to software programming errors and vulnerabilities.

Patch management—An area of systems management that involves acquiring, testing and installing multiple patches (code changes) to an administered computer system in order to maintain up-to-date software and often to address security risk. Patch management tasks include the following: maintaining current knowledge of available patches; deciding what patches are appropriate for particular systems; ensuring that patches are installed properly; testing systems after installation; and documenting all associate procedures, such as specific configurations required. A number of products are available to automate patch management tasks. Patches are sometimes ineffective and can sometimes cause more problems than they fix. Patch management experts suggest that system administrators take simple steps to avoid problems, such as performing backups and testing patches on noncritical systems prior to installations. Patch management can be viewed as part of change management.

Payload—The section of fundamental data in a transmission. In malicious software this refers to the section containing the harmful data/code.

Penetration testing—A live test of the effectiveness of security defenses through mimicking the actions of real-life attackers.

Personal identification number (PIN)—A type of password (i.e., a secret number assigned to an individual) that, in conjunction with some means of identifying the individual, serves to verify the authenticity of the individual. PINs have been adopted by financial institutions as the primary means of verifying customers in an electronic funds transfer (EFT) system.

Phishing—This is a type of electronic mail (email) attack that attempts to convince a user that the originator is genuine, but with the intention of obtaining information for use in social engineering. Phishing attacks may take the form of masquerading as a lottery organization advising the recipient or the user's bank of a large win; in either case, the intent is to obtain account and personal identification number (PIN) details. Alternative attacks may seek to obtain apparently innocuous business information, which may be used in another form of active attack.

Plain old telephone service (POTS)—A wired telecommunications system.

Platform as a Service (PaaS)—Offers the capability to deploy onto the cloud infrastructure customer- created or -acquired applications that are created using programming languages and tools supported by the provider.

Appendix B—Glossary

Policy—Generally, a document that records a high-level principle or course of action that has been decided on. The intended purpose is to influence and guide both present and future decision making to be in line with the philosophy, objectives and strategic plans established by the enterprise's management teams.

In addition to policy content, policies need to describe the consequences of failing to comply with the policy, the means for handling exceptions, and the manner in which compliance with the policy will be checked and measured.

Port (port number)—A process or application- specific software element serving as a communication end point for the Transport Layer IP protocols (UDP and TCP).

Port scanning—The act of probing a system to identify open ports.

Prime number—A natural number greater than 1 that can only be divided by 1 and itself.

Principle of least privilege/access—Controls used to allow the least privilege access needed to complete a task.

Privacy—Freedom from unauthorized intrusion or disclosure of information about an individual. probe—Inspect a network or system to find weak spots.

Procedure—A document containing a detailed description of the steps necessary to perform specific operations in conformance with applicable standards. Procedures are defined as part of processes.

Protocol—The rules by which a network operates and controls the flow and priority of transmissions.

Proxy server—A server that acts on behalf of a user. Typical proxies accept a connection from a user, make a decision as to whether the user or client IP address is permitted to use the proxy, perhaps perform additional authentication, and complete a connection to a remote destination on behalf of the user.

Public key encryption—A cryptographic system that uses two keys: one is a public key, which is known to everyone, and the second is a private or secret key, which is only known to the recipient of the message. See also Asymmetric Key.

Public key infrastructure (PKI)—A series of processes and technologies for the association of cryptographic keys with the entity to whom those keys were issued.

Public switched telephone network (PSTN)—A communications system that sets up a dedicated channel (or circuit) between two points for the duration of the transmission.

R

Ransomware—Malware that restricts access to the compromised systems until a ransom demand is satisfied.

Reciprocal agreement—Emergency processing agreement between two or more enterprises with similar equipment or applications. Typically, participants of a reciprocal agreement promise to provide processing time to each other when an emergency arises.

Recovery—The phase in the incident response plan that ensures that affected systems or services are restored to a condition specified in the service delivery objectives (SDOs) or business continuity plan (BCP).

Recovery action—Execution of a response or task according to a written procedure.

Recovery point objective (RPO)—Determined based on the acceptable data loss in case of a disruption of operations. It indicates the earliest point in time that is acceptable to recover the data. The RPO effectively quantifies the permissible amount of data loss in case of interruption.

Recovery time objective (RTO)—The amount of time allowed for the recovery of a business function or resource after a disaster occurs.

Redundant site—A recovery strategy involving the duplication of key IT components, including data or other key business processes, whereby fast recovery can take place.

Registered ports—Registered ports--1024 through 49151: Listed by the IANA and on most systems can be used by ordinary user processes or programs executed by ordinary users.

Registration authority (RA)—The individual institution that validates an entity's proof of identity and ownership of a key pair.

Regulation—Rules or laws defined and enforced by an authority to regulate conduct.

Regulatory requirements—Rules or laws that regulate conduct and that the enterprise must obey to become compliant.

Remediation—After vulnerabilities are identified and assessed, appropriate remediation can take place to mitigate or eliminate the vulnerability.

Remote access (RAS)—Refers to any combination of hardware and software to enable the remote access to tools or information that typically reside on a network of IT devices.

Originally coined by Microsoft when referring to their built-in NT remote access tools, RAS was a service provided by Windows NT which allowed most of the services that would be available on a network to be accessed over a modem link. Over the years, many vendors have provided both hardware and software solutions to gain remote access to various types of networked information. In fact, most modern routers include a basic RAS capability that can be enabled for any dial-up interface.

Removable media—Any type of storage device that can be removed from the system while is running.

Repeaters—A physical layer device that regenerates and propagates electrical signals between two network segments. Repeaters receive signals from one network segment and amplify (regenerate) the signal to compensate for signals (analog or digital) distorted by transmission loss due to reduction of signal strength during transmission (i.e., attenuation).

Replay—The ability to copy a message or stream of messages between two parties and replay (retransmit) them to one or more of the parties.

Residual risk—The remaining risk after management has implemented a risk response.

Resilience—The ability of a system or network to resist failure or to recover quickly from any disruption, usually with minimal recognizable effect.

Return on investment (ROI)—A measure of operating performance and efficiency, computed in its simplest form by dividing net income by the total investment over the period being considered.

Return-oriented programming attacks—An exploit technique in which the attacker uses control of the call stack to indirectly execute cherry-picked machine instructions immediately prior to the return instruction in subroutines within the existing program code.

Risk—The combination of the probability of an event and its consequence (ISO/IEC 73).

Risk acceptance—If the risk is within the enterprise's risk tolerance or if the cost of otherwise mitigating the risk is higher than the potential loss, the enterprise can assume the risk and absorb any losses.

Appendix B—Glossary

Risk assessment—A process used to identify and evaluate risk and its potential effects. Risk assessments are used to identify those items or areas that present the highest risk, vulnerability or exposure to the enterprise for inclusion in the IS annual audit plan. Risk assessments are also used to manage the project delivery and project benefit risk.

Risk avoidance—The process for systematically avoiding risk, constituting one approach to managing risk.

Risk management—The coordinated activities to direct and control an enterprise with regard to risk. In the International Standard, the term "control" is used as a synonym for "measure." (ISO/IEC Guide 73:2002)

One of the governance objectives. Entails recognizing risk; assessing the impact and likelihood of that risk; and developing strategies, such as avoiding the risk, reducing the negative effect of the risk and/or transferring the risk, to manage it within the context of the enterprise's risk appetite. (COBIT 5)

Risk mitigation—The management of risk through the use of countermeasures and controls.

Risk reduction—The implementation of controls or countermeasures to reduce the likelihood or impact of a risk to a level within the organization's risk tolerance.

Risk tolerance—The acceptable level of variation that management is willing to allow for any particular risk as the enterprise pursues its objectives.

Risk transfer—The process of assigning risk to another enterprise, usually through the purchase of an insurance policy or by outsourcing the service.

Risk treatment—The process of selection and implementation of measures to modify risk (ISO/IEC Guide 73:2002).

Root cause analysis—A process of diagnosis to establish the origins of events, which can be used for learning from consequences, typically from errors and problems.

Rootkit—A software suite designed to aid an intruder in gaining unauthorized administrative access to a computer system.

Router—A networking device that can send (route) data packets from one local area network (LAN) or wide area network (WAN) to another, based on addressing at the network layer (Layer 3) in the open systems interconnection (OSI) model. Networks connected by routers can use different or similar networking protocols. Routers usually are capable of filtering packets based on parameters, such as source addresses, destination addresses, protocol and network applications (ports).

RSA—A public key cryptosystem developed by R. Rivest, A. Shamir and L. Adleman used for both encryption and digital signatures. The RSA has two different keys, the public encryption key and the secret decryption key. The strength of the RSA depends on the difficulty of the prime number factorization. For applications with high-level security, the number of the decryption key bits should be greater than 512 bits.

S

Safeguard—A practice, procedure or mechanism that reduces risk.

Secure Electronic Transaction (SET)—A standard that will ensure that credit card and associated payment order information travels safely and securely between the various involved parties on the Internet.

Secure Multipurpose Internet Mail Extensions (S/MIME)—Provides cryptographic security services for electronic messaging applications: authentication, message integrity and nonrepudiation of origin (using digital signatures) and privacy and data security (using encryption) to provide a consistent way to send and receive MIME data (RFC 2311).

Secure Socket layer (SSL)—A protocol that is used to transmit private documents through the Internet. The SSL protocol uses a private key to encrypt the data that are to be transferred through the SSL connection.

Secure hypertext transfer protocol (HTTPS)— An application layer protocol, HTTPS transmits individual messages or pages securely between a web client and server by establishing an SSL-type connection.

Secure Shell (SSH)—Network protocol that uses cryptography to secure communication, remote command line login and remote command execution between two networked computers.

Security as a Service (SECaaS)—The next generation of managed security services dedicated to the delivery, over the Internet, of specialized information-security services.

Security metrics—A standard of measurement used in management of security-related activities.

Security perimeter—The boundary that defines the area of security concern and security policy coverage.

Segmentation—Network segmentation is the process of logically grouping network assets, resources, and applications together into compartmentalized areas that have no trust of each other.

Segregation/separation of duties (SoD)—A basic internal control that prevents or detects errors and irregularities by assigning to separate individuals the responsibility for initiating and recording transactions and for the custody of assets. Segregation/ separation of duties is commonly used in large IT organizations so that no single person is in a position to introduce fraudulent or malicious code without detection.

Sensitivity—A measure of the impact that improper disclosure of information may have on an enterprise.

Service delivery objective (SDO)—Directly related to the business needs, SDO is the level of services to be reached during the alternate process mode until the normal situation is restored.

Service level agreement (SLA)—An agreement, preferably documented, between a service provider and the customer(s)/user(s) that defines minimum performance targets for a service and how they will be measured.

Simple mail transfer protocol (SMTP)—The standard electronic mail (email) protocol on the Internet.

Single factor authentication (SFA)—Authentication process that requires only the user ID and password to grant access.

Smart card—A small electronic device that contains electronic memory, and possibly an embedded integrated circuit. Smart cards can be used for a number of purposes including the storage of digital certificates or digital cash, or they can be used as a token to authenticate users.

Sniffing—The process by which data traversing a network are captured or monitored.

Social engineering—An attack based on deceiving users or administrators at the target site into revealing confidential or sensitive information.

Software as a Service (SaaS)—Offers the capability to use the provider's applications running on cloud infrastructure. The applications are accessible from various client devices through a thin client interface such as a web browser (e.g., web-based email).

Source routing specification—A transmission technique where the sender of a packet can specify the route that packet should follow through the network.

Spam—Computer-generated messages sent as unsolicited advertising.

Spear phishing—An attack where social engineering techniques are used to masquerade as a trusted party to obtain important information such as passwords from the victim.

Spoofing—Faking the sending address of a transmission in order to gain illegal entry into a secure system.

Spyware—Software whose purpose is to monitor a computer user's actions (e.g., websites visited) and report these actions to a third party, without the informed consent of that machine's owner or legitimate user. A particularly malicious form of spyware is software that monitors keystrokes to obtain passwords or otherwise gathers sensitive information such as credit card numbers, which it then transmits to a malicious third party. The term has also come to refer more broadly to software that subverts the computer's operation for the benefit of a third party.

SQL injection—Results from failure of the application to appropriately validate input. When specially crafted user-controlled input consisting of SQL syntax is used without proper validation as part of SQL queries, it is possible to glean information from the database in ways not envisaged during application design. (MITRE)

Stateful inspection—A firewall architecture that tracks each connection traversing all interfaces of the firewall and makes sure they are valid.

Statutory requirements—Laws created by government institutions.

Supervisory control and data acquisition (SCADA)—Systems used to control and monitor industrial and manufacturing processes, and utility facilities.

Switches—Typically associated as a data link layer device, switches enable local area network (LAN) segments to be created and interconnected, which has the added benefit of reducing collision domains in Ethernet-based networks.

Symmetric key encryption—System in which a different key (or set of keys) is used by each pair of trading partners to ensure that no one else can read their messages. The same key is used for encryption and decryption. See also Private Key Cryptosystem.

System development lifecycle (SDLC)—The phases deployed in the development or acquisition of a software system. SDLC is an approach used to plan, design, develop, test and implement an application system or a major modification to an application system. Typical phases of SDLC include the feasibility study, requirements study, requirements definition, detailed design, programming, testing, installation and postimplementation review, but not the service delivery or benefits realization activities.

System hardening—A process to eliminate as much security risk as possible by removing all nonessential software programs, protocols, services and utilities from the system.

T

Tangible asset—Any assets that has physical form.

Target—Person or asset selected as the aim of an attack.

Telnet—Network protocol used to enable remote access to a server computer. Commands typed are run on the remote server.

Threat—Anything (e.g., object, substance, human) that is capable of acting against an asset in a manner that can result in harm. A potential cause of an unwanted incident (ISO/IEC 13335).

Threat agent—Methods and things used to exploit a vulnerability. Examples include determination, capability, motive and resources.

Threat analysis/assessment—An evaluation of the type, scope and nature of events or actions that can result in adverse consequences; identification of the threats that exist against enterprise assets. The threat analysis usually defines the level of threat and the likelihood of it materializing.

Threat event—Any event during which a threat element/actor acts against an asset in a manner that has the potential to directly result in harm.

Threat vector—The path or route used by the adversary to gain access to the target.

Time lines—Chronological graphs where events related to an incident can be mapped to look for relationships in complex cases. Time lines can provide simplified visualization for presentation to management and other nontechnical audiences.

Token—A device that is used to authenticate a user, typically in addition to a username and password. A token is usually a device the size of a credit card that displays a pseudo random number that changes every few minutes.

Topology—The physical layout of how computers are linked together. Examples of topology include ring, star and bus.

Total cost of ownership (TCO)—Includes the original cost of the computer plus the cost of: software, hardware and software upgrades, maintenance, technical support, training, and certain activities performed by users.

Transmission control protocol (TCP)—A connection-based Internet protocol that supports reliable data transfer connections. Packet data are verified using checksums and retransmitted if they are missing or corrupted. The application plays no part in validating the transfer.

Transmission control protocol/Internet protocol (TCP/IP)—Provides the basis for the Internet; a set of communication protocols that encompass media access, packet transport, session communication, file transfer, electronic mail (email), terminal emulation, remote file access and network management.

Transport layer security (TLS)—A protocol that provides communications privacy over the Internet. The protocol allows client-server applications to communicate in a way that is designed to prevent eavesdropping, tampering, or message forgery (RFC 2246).

Transport Layer Security (TLS) is composed of two layers: the TLS Record Protocol and the TLS Handshake Protocol. The TLS Record Protocol provides connection security with some encryption method such as the Data Encryption Standard (DES). The TLS Record Protocol can also be used without encryption. The TLS Handshake Protocol allows the server and client to authenticate each other and to negotiate an encryption algorithm and cryptographic keys before data is exchanged.

Triple DES (3DES)—A block cipher created from the Data Encryption Standard (DES) cipher by using it three times.

Trojan horse—Purposefully hidden malicious or damaging code within an authorized computer program. Unlike viruses, they do not replicate themselves, but they can be just as destructive to a single computer.

Tunnel—The paths that the encapsulated packets follow in an Internet virtual private network (VPN).

Tunnel mode—Used to protect traffic between different networks when traffic must travel through intermediate or untrusted networks. Tunnel mode encapsulates the entire IP packet with and AH or ESP header and an additional IP header.

Two-factor authentication—The use of two independent mechanisms for authentication, (e.g., requiring a smart card and a password) typically the combination of something you know, are or have.

U

Uncertainty—The difficulty of predicting an outcome due to limited knowledge of all components.

Uniform resource locator (URL)—The string of characters that form a web address.

User Datagram protocol (UDP)—A connectionless Internet protocol that is designed for network efficiency and speed at the expense of reliability. A data request by the client is served by sending packets without testing to verify whether they actually arrive at the destination, not whether they were corrupted in transit. It is up to the application to determine these factors and request retransmissions.

User interface impersonation—Can be a pop-up ad that impersonates a system dialog, an ad that impersonates a system warning, or an ad that impersonates an application user interface in a mobile device.

User mode—Used for the execution of normal system activities.

User provisioning—A process to create, modify, disable and delete user accounts and their profiles across IT infrastructure and business applications.

V

Value—The relative worth or importance of an investment for an enterprise, as perceived by its key stakeholders, expressed as total life cycle benefits net of related costs, adjusted for risk and (in the case of financial value) the time value of money.

Vertical defense in depth—Controls are placed at different system layers – hardware, operating system, application, database or user levels.

Virtual local area network (VLAP)—Logical segmentation of a LAN into different broadcast domains. A VLAN is set up by configuring ports on a switch, so devices attached to these ports may communicate as if they were attached to the same physical network segment, although the devices are located on different LAN segments. A VLAN is based on logical rather than physical connections.

Virtual private network (VPN)—A secure private network that uses the public telecommunications infrastructure to transmit data. In contrast to a much more expensive system of owned or leased lines that can only be used by one company, VPNs are used by enterprises for both extranets and wide areas of intranets. Using encryption and authentication, a VPN encrypts all data that pass between two Internet points, maintaining privacy and security.

Virtual private network (VPN) concentrator—A system used to establish VPN tunnels and handle large numbers of simultaneous connections. This system provides authentication, authorization and accounting services.

Virtualization—The process of adding a "guest application" and data onto a "virtual server," recognizing that the guest application will ultimately part company from this physical server.

Virus—A program with the ability to reproduce by modifying other programs to include a copy of itself. A virus may contain destructive code that can move into multiple programs, data files or devices on a system and spread through multiple systems in a network.

Virus signature file—The file of virus patterns that are compared with existing files to determine whether they are infected with a virus or worm.

Voice over Internet protocol (VOIP)—Also called IP Telephony, Internet Telephony and Broadband Phone, a technology that makes it possible to have a voice conversation over the Internet or over any dedicated Internet Protocol (IP) network instead of over dedicated voice transmission lines.

Volatile data—Data that changes frequently and can be lost when the system's power is shut down.

Vulnerability—A weakness in the design, implementation, operation or internal control of a process that could expose the system to adverse threats from threat events.

Vulnerability analysis/assessment—A process of identifying and classifying vulnerabilities.

Vulnerability scanning—An automated process to proactively identify security weaknesses in a network or individual system.

W

Warm site—Similar to a hot site but not fully equipped with all of the necessary hardware needed for recovery.

Web hosting—The business of providing the equipment and services required to host and maintain files for one or more websites and provide fast Internet connections to those sites. Most hosting is "shared," which means that websites of multiple companies are on the same server to share/reduce costs.

Web server—Using the client-server model and the World Wide Web's HyperText Transfer Protocol (HTTP), Web Server is a software program that serves web pages to users.

Well-known ports—0 through 1023: Controlled and assigned by the Internet Assigned Numbers Authority (IANA), and on most systems can be used only by system (or root) processes or by programs executed by privileged users. The assigned ports use the first portion of the possible port numbers. Initially, these assigned ports were in the range 0-255. Currently, the range for assigned ports managed by the IANA has been expanded to the range 0-1023.

Wide area network (WAN)—A computer network connecting different remote locations that may range from short distances, such as a floor or building, to extremely long transmissions that encompass a large region or several countries.

Wi-Fi protected access (WPA)—A class of systems used to secure wireless (Wi-Fi) computer networks. WPA was created in response to several serious weaknesses that researchers found in the previous system, Wired Equivalent Privacy (WEP). WPA implements the majority of the IEEE 802.11i standard, and was intended as an intermediate measure to take the place of WEP while 802.11i was prepared. WPA is designed to work with all wireless network interface cards, but not necessarily with first generation wireless access points. WPA2 implements the full standard, but will not work with some older network cards. Both provide good security with two significant issues. First, either WPA or WPA2 must be enabled and chosen in preference to WEP; WEP is usually presented as the first security choice in most installation instructions. Second, in the "personal" mode, the most likely choice for homes and small offices, a pass phrase is required that, for full security, must be longer than the typical six to eight character passwords users are taught to employ.

Wi-Fi protected access II (WPA2)—Wireless security protocol that supports 802.11i encryption standards to provide greater security. This protocol uses Advanced Encryption Standards (AES) and Temporal Key Integrity Protocol (TKIP) for stronger encryption.

Wired equivalent privacy (WEP)—A scheme that is part of the IEEE 802.11 wireless networking standard to secure IEEE 802.11 wireless networks (also known as Wi-Fi networks). Because a wireless network broadcasts messages using radio, it is particularly susceptible to eavesdropping. WEP was intended to provide comparable confidentiality to a traditional wired network (in particular, it does not protect users of the network from each other), hence the name. Several serious weaknesses were identified by cryptanalysts, and WEP was superseded by Wi-Fi Protected Access (WPA) in 2003, and then by the full IEEE 802.11i standard (also known as WPA2) in 2004. Despite the weaknesses, WEP provides a level of security that can deter casual snooping.

Wireless local area network (WLAN)—Two or more systems networked using a wireless distribution method.

Worm—A programmed network attack in which a self-replicating program does not attach itself to programs, but rather spreads independently of users' action.

Write blocker—A devices that allows the acquisition of information on a drive without creating the possibility of accidentally damaging the drive.

Write protect—The use of hardware or software to prevent data to be overwritten or deleted.

Z
Zero-day exploit—A vulnerability that is exploited before the software creator/vendor is even aware of its existence.

APPENDIX C—KNOWLEDGE CHECK ANSWERS

SECTION 1—KNOWLEDGE CHECK (PG. 21)

1. Three common controls used to protect the availability of information are:
   A. **redundancy, backups and access controls.**
   B. encryption, file permissions and access controls.
   C. access controls, logging and digital signatures.
   D. hashes, logging and backups.

2. Select all that apply. Governance has several goals, including:
   A. **providing strategic direction.**
   B. **ensuring that objectives are achieved.**
   C. **verifying that organizational resources are being used appropriately.**
   D. directing and monitoring security activities.
   E. **ascertaining whether risk is being managed properly.**

3. Choose three. According to the NIST cybersecurity framework, which of the following are considered key functions necessary for the protection of digital assets?
   A. Encrypt
   B. **Protect**
   C. Investigate
   D. **Recover**
   E. **Identify**

4. Which of the following is the best definition for cybersecurity?
   A. The process by which an organization manages cybersecurity risk to an acceptable level
   B. The protection of information from unauthorized access or disclosure
   C. The protection of paper documents, digital and intellectual property, and verbal or visual communications
   D. **Protecting information assets by addressing threats to information that is processed, stored or transported by internetworked information systems**

5. Which of the following cybersecurity roles is charged with the duty of managing incidents and remediation?
   A. Board of directors
   B. Executive committee
   C. **Cybersecurity management**
   D. Cybersecurity practitioners

SECTION 2—KNOWLEDGE CHECK (PG. 48)

1. The core duty of cybersecurity is to identify, mitigate and manage **cyberrisk** to an organization's digital assets.
2. A(n) **threat** is anything capable of acting against an asset in a manner that can cause harm.
3. A(n) **asset** is something of value worth protecting.
4. A(n) **vulnerability** is a weakness in the design, implementation, operation or internal controls in a process that could be exploited to violate the system security.
5. The path or route used to gain access to the target asset is known as a(n) **attack vector**.
6. In an attack, the container that delivers the exploit to the target is called a(n) **payload**.
7. **Policies** communicate required and prohibited activities and behaviors.
8. **Rootkit** is a class of malware that hides the existence of other malware by modifying the underlying operating system.
9. **Procedures** provide details on how to comply with policies and standards.
10. **Guidelines** provide general guidance and recommendations on what to do in particular circumstances.

Appendix C—Knowledge Check Answers

11. **Malware**, also called malicious code, is software designed to gain access to targeted computer systems, steal information or disrupt computer operations.
12. **Standards** are used to interpret policies in specific situations.
13. **Patches** are solutions to software programming and coding errors.
14. **Identity management** includes many components such as directory services, authentication and authorization services, and user management capabilities such as provisioning and deprovisioning.

SECTION 3—KNOWLEDGE CHECK (PG. 83)

1. Select all that apply. The Internet perimeter should:
   A. **detect and block traffic from infected internal end points.**
   B. **eliminate threats such as email spam, viruses and worms.**
   C. format, encrypt and compress data.
   D. **control user traffic bound toward the Internet.**
   E. **monitor internal and external network ports for rogue activity.**

2. The _____ layer of the OSI model ensures that data are transferred reliably in the correct sequence, and the _____ layer coordinates and manages user connections.
   A. Presentation, data link
   B. **Transport, session**
   C. Physical, application
   D. Data link, network

3. Choose three. There key benefits of the DMZ system are:
   A. DMZs are based on logical rather than physical connections.
   B. **an intruder must penetrate three separate devices.**
   C. **private network addresses are not disclosed to the Internet.**
   D. excellent performance and scalability as Internet usage grows.
   E. **internal systems do not have direct access to the Internet.**

4. Which of the following best states the role of encryption within an overall cybersecurity program?
   A. Encryption is the primary means of securing digital assets.
   B. Encryption depends upon shared secrets and is therefore an unreliable means of control.
   C. A program's encryption elements should be handled by a third-party cryptologist.
   D. **Encryption is an essential but incomplete form of access control.**

5. The number and types of layers needed for defense in depth are a function of:
   A. **asset value, criticality, reliability of each control and degree of exposure.**
   B. threat agents, governance, compliance and mobile device policy.
   C. network configuration, navigation controls, user interface and VPN traffic.
   D. isolation, segmentation, internal controls and external controls.

SECTION 4—KNOWLEDGE CHECK (PG. 118)

1. Put the steps of the penetration testing phase into the correct order.
D. Planning
B. Discovery
A. Attack
C. Reporting

2. System hardening should implement the principle of _____ or _____ .
   A. Governance, compliance
   B. **Least privilege, access control**
   C. Stateful inspection, remote access
   D. Vulnerability assessment, risk mitigation

3. Select all that apply. Which of the following are considered functional areas of network management as defined by ISO?
   A. **Accounting management**
   B. **Fault management**
   C. Firewall management
   D. **Performance management**
   E. **Security management**

4. Virtualization involves:
   A. the creation of a layer between physical and logical access controls.
   B. **multiple guests coexisting on the same server in isolation of one another.**
   C. simultaneous use of kernel mode and user mode.
   D. DNS interrogation, WHOIS queries and network sniffing.

5. Vulnerability management begins with an understanding of cybersecurity assets and their locations, which can be accomplished by:
   A. vulnerability scanning.
   B. penetration testing.
   C. **maintaining an asset inventory.**
   D. using command line tools.

SECTION 5—KNOWLEDGE CHECK (PG. 137)

1. Arrange the steps of the incident response process into the correct order.
   D. Preparation
   E. Detection and analysis
   B. Investigation
   A. Mitigation and recovery
   C. Postincident analysis

2. Which element of an incident response plan involves obtaining and preserving evidence?
   A. Preparation
   B. Identification
   C. **Containment**
   D. Eradication

3. Select three. The chain of custody contains information regarding:
   A. Disaster recovery objectives, resources and personnel.
   B. **Who had access to the evidence, in chronological order.**
   C. Labor, union and privacy regulations.
   D. **Proof that the analysis is based on copies identical to the original evidence.**
   E. **The procedures followed in working with the evidence.**

Appendix C—Knowledge Check Answers

4. NIST defines a(n) as a "violation or imminent threat of violation of computer security policies, acceptable use policies, or standard security practices."
   A. Disaster
   B. Event
   C. Threat
   D. Incident

5. Select all that apply. A business impact analysis (BIA) should identify:
   A. the circumstances under which a disaster should be declared.
   B. the estimated probability of the identified threats actually occurring.
   C. the efficiency and effectiveness of existing risk mitigation controls.
   D. a list of potential vulnerabilities, dangers and/or threats.
   E. which types of data backups (full, incremental and differential) will be used.

SECTION 6—KNOWLEDGE CHECK (PG. 161)

1. _____ is defined as "a model for enabling convenient, on-demand network access to a shared pool of configurable resources (e.g., networks, servers, storage, applications and services) that can be rapidly provisioned and released with minimal management or service provider interaction."
   A. Software as a Service (SaaS)
   B. Cloud computing
   C. Big data
   D. Platform as a Service (PaaS)

2. Select all that apply. Which of the following statements about advanced persistent threats (APTs) are true?
   A. APTs typically originate from sources such as organized crime groups, activists or governments.
   B. APTs use obfuscation techniques that help them remain undiscovered for months or even years.
   C. APTs are often long-term, multi-phase projects with a focus on reconnaissance.
   D. The APT attack cycle begins with target penetration and collection of sensitive information.
   E. Although they are often associated with APTs, intelligence agencies are rarely the perpetrators of APT attacks.

3. Which of the following are benefits to BYOD?
   A. Acceptable Use Policy is easier to implement.
   B. Costs shift to the user.
   C. Worker satisfaction increases.
   D. Security risk is known to the user.

4. Choose three. Which types of risk are typically associated with mobile devices?
   A. Organizational risk
   B. Compliance risk
   C. Technical risk
   D. Physical risk
   E. Transactional risk

5. Which three elements of the current threat landscape have provided increased levels of access and connectivity, and, therefore, increased opportunities for cybercrime?
   A. Text messaging, Bluetooth technology and SIM cards
   B. Web applications, botnets and primary malware
   C. Financial gains, intellectual property and politics
   D. Cloud computing, social media and mobile computing